T0331192

Architectured
Multifunctional
Materials

MATERIALS RESEARCH SOCIETY
SYMPOSIUM PROCEEDINGS VOLUME 1188

Architectured Multifunctional Materials

Symposium held April 14–16, San Francisco, California, U.S.A.

EDITORS:

Yves J.M. Brechet
Grenoble Institute of Technology
St. Martin d'Heres, France

J. David Embury
McMaster University
Ontario, Canada

Patrick R. Onck
University of Groningen
Groningen, The Netherlands

Materials Research Society
Warrendale, Pennsylvania

CAMBRIDGE
UNIVERSITY PRESS

University Printing House, Cambridge CB2 8BS, United Kingdom

One Liberty Plaza, 20th Floor, New York, NY 10006, USA

477 Williamstown Road, Port Melbourne, VIC 3207, Australia

314-321, 3rd Floor, Plot 3, Splendor Forum, Jasola District Centre, New Delhi - 110025, India

79 Anson Road, #06-04/06, Singapore 079906

Cambridge University Press is part of the University of Cambridge.

It furthers the University's mission by disseminating knowledge in the pursuit of education, learning and research at the highest international levels of excellence.

www.cambridge.org
Information on this title: www.cambridge.org/9781605111612

Materials Research Society
506 Keystone Drive, Warrendale, PA 15086
http://www.mrs.org

© Materials Research Society 2009

First published 2009
First paperback edition 2012

Single article reprints from this publication are available through University Microfilms Inc., 300 North Zeeb Road, Ann Arbor, MI 48106

CODEN: MRSPDH

A catalogue r5ecord for this publication is available from the British Library

ISBN 978-1-605-11161-2 Hardback
ISBN 978-1-107-40817-3 Paperback

BASIC CONCEPTS IN
ARCHITECTURED MATERIALS

CELLULAR AND FIBROUS MATERIALS

*Invited Paper

PROCESSING CHALLENGES

TOWARDS STRUCTURES

*Invited Paper

vi

MULTIFUNCTIONAL MATERIALS

*Invited Paper

PREFACE

Multifunctional requirements are becoming the rule in terms of design. Weight saving, safety and energy management are societal requirements that cannot be ignored, and an efficient use of matter requires that several functions can be obtained in the same component. While the last decades have been driven by the development of new materials, or by the appropriate choice of materials for a given application, this new and demanding engineering approach triggers the development of "tailored materials."

Very often, contradictory requirements cannot be met by single materials. The development of new "materials," in a more general sense, able to combine otherwise contradictory properties, cannot be achieved by classical "alloy design" or "combinatorial polymer chemistry." Various innovative strategies are possible: either combining different materials (such as multilayers), or playing with material architectures (such as foams or truss lattices), or developing microstructural gradients. These strategies open a whole new range of materials and properties, where structural requirements and functional properties can be combined. They also reveal new challenges such as implementing new processes, developing appropriate constitutive equations, engineering interfaces, developing and modeling bio-inspired hierarchical structures, and promoting design methods to deal systematically with optimizing this new class of materials. These strategies are especially suited for a "materials by design" approach where material combinations, microstructural gradients and multiscale architectures are optimized to meet a complex set of requirements, possibly leading to a combination of properties that is otherwise impossible to reach. This is a whole new field of research and engineering that is opened by these new strategies.

Symposium LL, "Architectured Multifunctional Materials," held April 14–16 at the 2009 MRS Spring Meeting in San Francisco, California, the first of its kind on this very general topic, was an attempt to bring together physicists, chemists, materials scientists and mechanicians dealing with a wide variety of fields of application. It is hopeless, in such a wide and rapidly evolving field, to aim at an exhaustive description. More to the point is to highlight the spirit of the new approach so that other researchers can bring in their own competences, their own creativity. Both functional and structural properties are on the agenda, materials are metals, polymers and ceramics, processing routes are from all the possible states of matter, solid, liquid, gaseous. Architectures under consideration span all scales of condensed matter, from the nanoscale to the millimeter. The common point is the impossibility to fulfill the required properties with a single unstructured material; the spirit is to design the material for the requirement. Seventy contributions were presented at the symposium, a selection of 29 proceedings papers in this volume give a glance at the variety of approaches, and at the future developments.

The organization of the proceedings reflects this variety. The first section gathers contributions on the basic concepts, materials combinations, optimized geometries, and hierarchical structures, with an insight in the possibility to seek inspiration from natural materials. The "ancestors" of architectured materials, namely fibrous and cellular structures will illustrate the concepts. A third group of papers deal with the new processes involved and required by complex materials and architecture combinations. Architectured materials are somewhere between microstructure and structure: a whole range of possibilities is open at the "millimeter scale" where geometry can be more easily controlled; this will be the leitmotiv of the "toward structures" section. Finally, the last group of papers illustrate the driving force behind these new strategies in materials developments: looking for "multifunctional materials."

There is one contribution missing from this volume, that by Professor Tony Evans on "Geometrical Aspects of Architectured Materials Design." Sadly, Professor Evans died during the preparation of this volume and was unable to contribute the written version of his paper. He was a man of great dedication and courage and we would like to dedicate this volume to his memory. A small tribute to a man who was an inspiration and mentor to many researchers and who was indeed one of the pioneers in this emerging area of materials research.

Yves J.M. Brechet
J. David Embury
Patrick R. Onck

June 2009

INTRODUCTION

M.F. Ashby

University of Cambridge, Engineering Department,
Trumpington Street, CB21PZ, Cambridge, UK

"Architecture: the Art and Science of structural design." It is a definition that works well here. When the population of Cro-Magnon man first exceeded the capacity of caves and other natural habitats, the need to build structures was forced upon them, and architecture, one of mankind's oldest creative activities, was born. It provided shelter, protection from predators and, later, storage for grain and security for livestock. It was, in a word, *multifunctional*.

Today, three million years later, we have multifunctional architectured *materials*. The papers in this volume echo the long-past history in more than one way. There is the Art – the ingenuity in devising processing paths for controlling the scale and connectivity of foam-like cellular structures, for welding tiny hollow sphere into close-packed arrays and for binding fibers into brush-like bundles. There is Art, too, in the elegant microscopic and tomographic images by which they are characterized. And there is the Science – the physical testing, the physical modeling and the digital simulation that allow properties of micro-architectured structures to be understood and predicted.

Architectured materials had, of course, evolved millions of years before *homo sapiens* appeared on earth. The multilevel structures of wood, bone, coral and shell exploit the advantages offered by cellular microstructure, composite reinforcement and sandwich design. They provide inspiration (even, sometimes, templates) for the design of architectured materials of today. The external or internal skeleton of animals has to provide stiffness and strength, yet it must also allow articulation if the creature is to move. Nature solves this problem by creating what, in contemporary terminology, are called tensegrity structures: rigid interlocking vertebra or scales-like segments, pulled together at joints by tendon and muscle to allow relative motion. The key feature, preventing dislocation, is the topological interlock of the joints. This has inspired a new set of architectured structures of a different kind, made up of discrete small but interlocking blocks, held together by surrounding tensile "ligaments" that promise to expand the use of brittle ceramics in load bearing structures.

Multifunctionality is a much misused word, easy to claim – even the most ordinary material can, after all, be used in more than one way – and can, or could, perform more than one function. To justify the claim of "multifunctionality" beyond the ordinary, an architectured material must either allow two properties that are usually coupled to be decoupled, allowing independent optimization of each, or it must perform useful functions in more than one realm, simultaneously providing mechanical, thermal, electrical or optical functions that are beyond the scope of simple homogeneous materials. The papers of this volume provide examples of both. They include materials with uniquely-tunable values of stiffness, strength and thermal expansion, and materials with independently adjustable strength, weight and acoustic damping. But despite the long history of its antecedents, this is a field in its infancy, and this symposium a report of work in progress, exciting work; with the promise of more excitement to come.

MATERIALS RESEARCH SOCIETY SYMPOSIUM PROCEEDINGS

Volume 1153 — Amorphous and Polycrystalline Thin-Film Silicon Science and Technology — 2009, A. Flewitt, Q. Wang, J. Hou, S. Uchikoga, A. Nathan, 2009, ISBN 978-1-60511-126-1

Volume 1154 — Concepts in Molecular and Organic Electronics, N. Koch, E. Zojer, S.-W. Hla, X. Zhu, 2009, ISBN 978-1-60511-127-8

Volume 1155 — CMOS Gate-Stack Scaling — Materials, Interfaces and Reliability Implications, J. Butterbaugh, A. Demkov, R. Harris, W. Rachmady, B. Taylor, 2009, ISBN 978-1-60511-128-5

Volume 1156— Materials, Processes and Reliability for Advanced Interconnects for Micro- and Nanoelectronics — 2009, M. Gall, A. Grill, F. Iacopi, J. Koike, T. Usui, 2009, ISBN 978-1-60511-129-2

Volume 1157 — Science and Technology of Chemical Mechanical Planarization (CMP), A. Kumar, C.F. Higgs III, C.S. Korach, S. Balakumar, 2009, ISBN 978-1-60511-130-8

Volume 1158E —Packaging, Chip-Package Interactions and Solder Materials Challenges, P.A. Kohl, P.S. Ho, P. Thompson, R. Aschenbrenner, 2009, ISBN 978-1-60511-131-5

Volume 1159E —High-Throughput Synthesis and Measurement Methods for Rapid Optimization and Discovery of Advanced Materials, M.L. Green, I. Takeuchi, T. Chiang, J. Paul, 2009, ISBN 978-1-60511-132-2

Volume 1160 — Materials and Physics for Nonvolatile Memories, Y. Fujisaki, R. Waser, T. Li, C. Bonafos, 2009, ISBN 978-1-60511-133-9

Volume 1161E —Engineered Multiferroics — Magnetoelectric Interactions, Sensors and Devices, G. Srinivasan, M.I. Bichurin, S. Priya, N.X. Sun, 2009, ISBN 978-1-60511-134-6

Volume 1162E —High-Temperature Photonic Structures, V. Shklover, S.-Y. Lin, R. Biswas, E. Johnson, 2009, ISBN 978-1-60511-135-3

Volume 1163E —Materials Research for Terahertz Technology Development, C.E. Stutz, D. Ritchie, P. Schunemann, J. Deibel, 2009, ISBN 978-1-60511-136-0

Volume 1164 — Nuclear Radiation Detection Materials — 2009, D.L. Perry, A. Burger, L. Franks, K. Yasuda, M. Fiederle, 2009, ISBN 978-1-60511-137-7

Volume 1165 — Thin-Film Compound Semiconductor Photovoltaics — 2009, A. Yamada, C. Heske, M. Contreras, M. Igalson, S.J.C. Irvine, 2009, ISBN 978-1-60511-138-4

Volume 1166 — Materials and Devices for Thermal-to-Electric Energy Conversion, J. Yang, G.S. Nolas, K. Koumoto, Y. Grin, 2009, ISBN 978-1-60511-139-1

Volume 1167 — Compound Semiconductors for Energy Applications and Environmental Sustainability, F. Shahedipour-Sandvik, E.F. Schubert, L.D. Bell, V. Tilak, A.W. Bett, 2009, ISBN 978-1-60511-140-7

Volume 1168E —Three-Dimensional Architectures for Energy Generation and Storage, B. Dunn, G. Li, J.W. Long, E. Yablonovitch, 2009, ISBN 978-1-60511-141-4

Volume 1169E —Materials Science of Water Purification, Y. Cohen, 2009, ISBN 978-1-60511-142-1

Volume 1170E —Materials for Renewable Energy at the Society and Technology Nexus, R.T. Collins, 2009, ISBN 978-1-60511-143-8

Volume 1171E —Materials in Photocatalysis and Photoelectrochemistry for Environmental Applications and H_2 Generation, A. Braun, P.A. Alivisatos, E. Figgemeier, J.A. Turner, J. Ye, E.A. Chandler, 2009, ISBN 978-1-60511-144-5

Volume 1172E —Nanoscale Heat Transport — From Fundamentals to Devices, R. Venkatasubramanian, 2009, ISBN 978-1-60511-145-2

Volume 1173E —Electofluidic Materials and Applications — Micro/Biofluidics, Electowetting and Electrospinning, A. Steckl, Y. Nemirovsky, A. Singh, W.-C. Tian, 2009, ISBN 978-1-60511-146-9

Volume 1174 — Functional Metal-Oxide Nanostructures, J. Wu, W. Han, A. Janotti, H.-C. Kim, 2009, ISBN 978-1-60511-147-6

MATERIALS RESEARCH SOCIETY SYMPOSIUM PROCEEDINGS

Prior Materials Research Society Symposium Proceedings available by contacting Materials Research Society

Basic Concepts in
Architectured Materials

Mater. Res. Soc. Symp. Proc. Vol. 1188 © 2009 Materials Research Society 1188-LL01-01

Mechanical Principles of a Self-Similar Hierarchical Structure

Huajian Gao
Division of Engineering, Brown University, Providence, RI, 02912, USA

ABSTRACT
Natural materials such as bone, shell, tendon and the attachment system of gecko exhibit multi-scale hierarchical structures. Here we summarize some recent studies on an idealized self-similar hierarchical model of bone and bone-like materials, and discuss mechanical principles of self-similar hierarchy, in particular to show how the characteristic length, aspect ratio and density at each hierarchical level can be selected to achieve flaw tolerance and superior stiffness and toughness across scale.
Tel.: (401) 863-2626; Email address: Huajian_Gao@Brown.edu

INTRODUCTION

Multi-level structural hierarchy can be observed in many biological systems including bone [1-6] and attachment pads of gecko [7-9]. In fact, structural hierarchy is a rule of nature. Hierarchical structures can be observed in all biosystems from chromosome, protein, cell, tissue, organism, to ecosystems. What are the roles and principles of structural hierarchy? What determine the size scales and other geometrical factors in a hierarchical material? These questions should be of general interest to both engineering and biological systems.

Recent studies on biological materials have shown that the characteristic size at each level of structural hierarchy may have been selected to ensure tolerance of material/structural flaws. For example, it has been demonstrated [10,11] that, due to their nanoscale characteristic size, the mineral bits in bone and bone-like materials tend to fail not by propagation of pre-existing cracks but by uniform rupture at the limiting strength of the material. For biological adhesion systems [7,8,12], similar transition from crack-like failure to uniform rupture has also been discussed [13]; the adhesion strength is affected not only by the size but also by the shape of the contacting surfaces: the smaller the size, the less important the shape, and shape-insensitive optimal adhesion was found to become possible when the structural size is reduced to below a critical length around 100 nm for van der Waals adhesion [14].

In this paper, we summarize some recent studies on an idealized self-similar hierarchical model mimicking the structure of bone [6,15]. It is known that bone and bone-like materials (Fig. 1) exhibit hierarchical structures over many length scales. For example, sea shells have 2 to 3 levels of lamellar structure [1,2,16-18], while vertebral bone has 7 levels of structural hierarchy [2, 19-22] (Fig. 2). Although the higher level structures of bone and bone-like materials show great complexity and variations, they exhibit a generic nanostructure (Fig. 2) at the most elementary level of structural hierarchy consisting of nanometer sized hard mineral

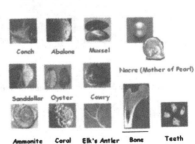

Figure 1. Bone and bone-like materials.

crystals arranged in a parallel staggered pattern in a soft protein matrix [3,5,10]. The nanostructure of tooth enamel shows needle-like (15–20nm thick and 1000nm long) crystals embedded in a relatively small volume fraction of a soft protein matrix [23-25]. The nanostructure of dentin and bone consists of plate-like (2–4nm thick and up to 100nm long) crystals embedded in a collagen-rich protein matrix [19,20], with the volume ratio of mineral to matrix on the order of 1 to 2. Nacre is made of very high volume fraction of plate-like crystals (200–500nm thick and a few micrometers long) with a small amount of soft matrix in between [1,2,16-18,26]. Bone and nacre are constructed with basically the same type of nanostructure made of staggered plate-like hard inclusions in a soft matrix. This staggered nanostructure is primarily subjected to uniaxial loading, as shown in Fig. 2.

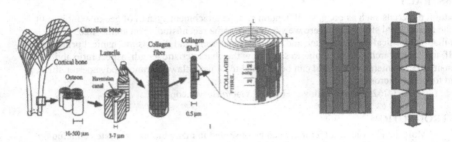

Figure 2. Hierarchical structures of bone and the generic nanostructure of bone-like materials [3, 10, 21].

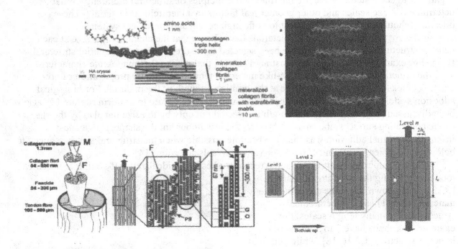

Figure 3. Self-similar structural hierarchy in bone and tendon [6,15,27,28].

As an example of hierarchical structures, in this paper we discuss an idealized hierarchical material with multiple levels of self-similar structures mimicking the staggered nanostructure of bone [6,15], as shown in Fig. 3. The resulting composite structure is still made of mineral and protein at some volume fraction, but the material is now distributed in a highly non-homogeneous way to form a hierarchical material with different properties at different length scales. In principle, this self-similar model can have arbitrary levels of hierarchy and at each level exhibits the same structure of slender hard plates arranged in a parallel staggered pattern in a soft matrix, similar to the nanostructure of bone. At each level of hierarchy, three mechanical principles will be required to determine three geometrical parameters: the width, length and volume fraction of hard inclusions.

MECHANICAL PRINCIPLES OF SELF-SIMILAR HIERARCHY

Principle I: Selection of characteristic length for uniform stress in hard particles at failure

Biological systems must be robust for survival. Therefore, the first principle of biocomposite is postulated to be that of flaw tolerance. Since the staggered biocomposite structure is primarily subjected to uniaxial tension, as shown in Fig. 2, the path of load transfer in the structure follows a tension-shear chain with the hard plates under tension and the soft matrix under shear. Analysis based on this tension-shear chain model [10, 29] showed that the integrity of biocomposites hinges upon the tensile strength of the hard particles. In order to keep the structure intact during deformation, the hard particles must be able to sustain large tensile load without fracture, while the protein/particle interface and the protein layer transfer load via shear and dissipate energy. The essence of flaw tolerance is to maximize the load carrying potential of the hard particles, which amounts to avoiding crack propagation until the material reaches its limiting strength. Theoretical estimation based on interatomic force laws shows that the theoretical strength is around $E/30$, where E is the Young's modulus. In reality, however, such high strength is rarely observed due to the inevitable presence of crack-like flaws which, under external loading, induce stress concentration near the tips of these flaws. As the external load reaches a critical value, the solid would fracture via crack propagation instead of simultaneous breaking of all bonds as assumed in the definition of *theoretical strength*. Under this circumstance, the load carrying capacity of the material is not utilized most efficiently since only a small fraction of material is maximally stressed at any instant of time during failure, leading to a

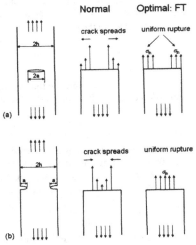

Figure 4. Illustration of flaw tolerance in a stretched strip particle with (a) a center crack or (b) two edge cracks. Under normal engineering circumstances, such cracked particle would fail by crack propagation at a critical applied load. However, as the strip width is reduced to below a critical value, the cracked particle would fail by uniform rupture near theoretical strength irrespective of the crack size. This state of material is termed flaw tolerance (FT).

5

much reduced "apparent" strength in contrast to the theoretical value. From the robustness point of view, an ideal scenario is to achieve the state of *flaw tolerance* [10,11] in which the material fails at the theoretical strength irrespective of the presence and size of cracks. Due to the random, unpredictable nature of crack-like flaws, it may seem at a first glance extremely difficult or impossible to eliminate stress concentration for large cracks so as to ensure uniform stress at material failure. However, theoretical arguments based on the well established concepts in fracture mechanics [6, 15] showed that this is actually possible with hierarchical material design.

The concept of uniform stress at failure can be demonstrated by a simple example involving an elastic strip particle containing a random internal or two edge cracks under tension (Fig. 4). Theoretical investigation of this problem based on the Dugdale cohesive model [11] showed that the cracked particle can indeed achieve uniform stress at failure for arbitrary crack size as long as the half-width h of the strip satisfies the following condition

$$h \leq h_{cr} = \frac{\Gamma E}{S^2},$$
(1)

where E is the Young's modulus, and S and Γ stand for the theoretical strength and fracture energy of the strip, respectively. For brittle materials, the fracture energy, which represents the amount of energy required for a unit increment of crack area, is usually taken to be twice of the surface energy, i.e., $\Gamma = 2\gamma$. Assuming no over-design of materials, we can just take the equality in Eq. (1) and write the condition of flaw tolerance as

$$\frac{\Gamma E}{S^2 h} = 1.$$
(2)

The principle of flaw tolerance thus yields a concrete equation to determine the characteristic dimension of the structure such that the stress in the structure is kept uniform even at the critical point of failure.

Principle II: Selection of aspect ratio for uniform stress in the matrix at failure

The next question is how the aspect ratio of the hard plates should be chosen. There have been two interesting considerations here. The first is to keep the stress in the matrix uniform up to failure. In a way, this is actually similar to Principle I. For this to happen, the length of the hard particle must not exceed a characteristic length defined as [30]

$$L_c = h\sqrt{\frac{2E\Theta_p(1-\varphi)}{\varphi\tau_p}}$$
(3)

where the subscript " p " stands for protein, τ_p is the shear yield strength of the protein matrix, Θ_p denotes the effective range of strain-to-failure of protein; φ denotes the volume fraction and E is the Young's modulus of hard particles. The concept of a characteristic length for stress transfer was first introduced via the classical shear lag model [31-33] and has played a key role in understanding mechanical properties of composites [34]. For a given length ℓ of hard particles, if we assume that the shear stress is equal to τ_p within the stress transfer zone and zero outside this zone at failure, we would expect that the maximum load transfer in the structure should satisfy the scaling law

$$\frac{P}{L_c\tau_p} = \begin{cases} \dfrac{\ell}{L_c} & \dfrac{\ell}{L_c} \le 1 \\ 1 & \dfrac{\ell}{L_c} \ge 1 \end{cases} \qquad (4)$$

Numerical simulations [30], as shown in Fig. 5, confirm that stress transfer in the staggered biocomposites is most efficient when the length of the hard particles is equal to the characteristic length L_c.

The second consideration is that, even when the shear stress in the matrix is uniform, as assumed in the tension-shear chain model, the aspect ratio of the particles could be further adjusted to ensure that the matrix, the particles and their interfaces reach their respective limiting strengths at the same time [6,10], i.e.

$$\tau_p = \tau_{int} = S/\rho \qquad (5)$$

Therefore, the aspect ratio of particles can be selected to ensure (1) that the stress in the matrix remains uniform even at the failure point and (2) that the soft matrix, the hard particles and their interfaces fail at the same time.

Similar concept of a characteristic length has been developed for optimal overlap length between two tropocollagen molecules [35], where it was shown that homogeneous shear is only possible when the overlap length is below the characteristic length and that fracture like slip pulses develop for larger overlapping lengths.

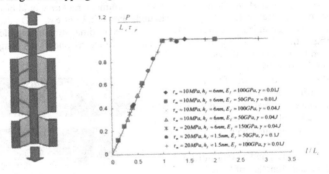

Figure 5. The scaling law of the normalized pulling force as a function of the normalized platelet length for the staggered composite. The simulation results under different combinations of parameters show excellent comparison with the theoretical prediction based on Eq. (4), which is shown as the solid line [30].

The principle of optimal stress transfer in the matrix thus provides a condition to determine the aspect ratio of particles as

$$\rho = \min\left(\frac{S}{\tau_p}, \sqrt{\frac{2E\Theta_p(1-\varphi)}{\varphi\tau_p}} \right) \qquad (6)$$

7

Principle III: Selection of volume fraction of hard particles for optimal balance between stiffness and toughness

After the characteristic length and aspect ratio of hard particles have been determined to ensure uniform stress of materials up to the failure point in both particles and matrix, the volume fraction of hard particles can be further selected to achieve the desired objective in mechanical properties. For bone, this objective is expected to be a combined stiffness and toughness. The toughness of the structure is expected to rise as the volume fraction of the hard material decreases. That is, the more soft material in the structure, the higher the toughness, although this may occur at the cost of compromising stiffness of the structure. For example, the toughness is expected to be of increasing importance for sea shells, bone and antelope horn, which correlates with their respective mineral contents of 95%, 45% and 35%, respectively. We will discuss shortly that the toughness of a hierarchical material at one level depends on the strength at the lower level, i.e. higher strength at the (n-1)th level results in higher toughness at the nth level. This is a unique property of hierarchical materials.

That bone and bone-like materials have been evolved to achieve a balance between toughness and stiffness should be apparent in their staggered structure. Studies based on genetic algorithms [36,6] have shown that the staggered nanostructure of bone could be interpreted as an evolutionary result with the objective to simultaneously optimize stiffness and toughness. This can be explained by comparing three rudimentary composite structures depicted in Fig. 6. If the objective is to optimize stiffness alone, the expected optimal structure would have the horizontal parallel strip configuration (the Voigt upper bound) shown in Figure 6(a). On the other hand, if the objective is to optimize toughness only, the expected optimal structure would be the vertical strip configuration shown in Figure 6(b) in which the soft matrix undergoes completely uniform deformation for maximum energy dissipation. In comparison with the staggered structure of Figure 6(c), the horizontal strip structure in Figure 6(a) is too brittle as the soft matrix does not have much chance to deform before the hard particles fail; on the other hand, the vertical strip structure in Figure 6(b) is too soft (as soft as the soft phase) and cannot fulfill the structural support function of bone.

BOTTOM-UP DESIGN OF SELF-SIMILAR HIERARCHY

We now apply the above principles to assemble a self-similar staggered hierarchical composite structure [6,15]. Consider a total of *N* hierarchical levels. At each level, the hard particles and the soft matrix play essentially the same roles, i.e. the slender particles provide structural rigidity while the soft matrix absorbs and dissipates mechanical energy [6,10,15]. For

Figure 6. Optimized hard–soft structures with different objective functions. (a) The columnar structure parallel to the direction of loading is expected if the objective is to optimize the stiffness alone. (b) The columnar structure perpendicular to the direction of loading results in uniform deformation in the soft matrix and is expected to be the optimal structure if the objective is to optimize the amount of energy absorption. (c) The staggered structure is expected to be optimal if the objectiveis to optimize both stiffness and toughness of the structure [6].

the self-similar structure, the same three principles, i.e. flaw tolerance, optimal stress transfer and optimal stiffness and toughness, are applied to each level of hierarchy following a bottom-up route. First, the characteristic size scale of the lowest level of structure is determined. Then the properties at the next higher level are calculated based on the current level of structures, and the characteristic size of the next level is determined by using the criterion of flaw tolerance. This iterative process is repeated until all N levels are determined [6,15].

At the lowest level, the flaw tolerance condition can be readily expressed in terms of the material constants of mineral as

$$\frac{\Gamma_0 E_0}{S_0^2 h_0} = \frac{2\gamma E_m}{\sigma_{th}^2 h_0} = 1, \tag{7}$$

where $E_0 = E_m$ is the Young's modulus of the mineral, γ is the surface energy and σ_{th} is the theoretical strength of mineral. According to Principle I, The characteristic size of the mineral particles is selected as

$$h_0 = \frac{2\gamma E_m}{\sigma_{th}^2}. \tag{8}$$

For bio-minerals, we take $\gamma = 1$ J/m^2, $E_m = 100$ GPa and $\sigma_{th} = E_m / 30$, and find

$$h_0 = 18 \text{ nm}.$$

This nanoscale size becomes the basis for designing structures at higher levels of hierarchy. Assuming that the structure of the n-th hierarchical level has been determined, the effective Young's modulus at the $(n+1)$-th hierarchical level E_{n+1} can be calculated as [6,10,15] as

$$\frac{1}{E_{n+1}} = \frac{4(1-\varphi_n)}{G_n^p \varphi_n^2 \rho_n^2} + \frac{1}{\varphi_n E_n}, \tag{9}$$

where G_n^p is the shear modulus of the soft matrix at the n-th level, and E_n, φ_n, ρ_n are Young's modulus, volume fraction and aspect ratio of the hard particles at the n-th level. Actually, the elastic properties of the hard particles at higher hierarchical levels are anisotropic. For simplicity, here we only consider the effective Young's modulus under uniaxial tension. In comparison, the Voigt upper bound of composite stiffness at the $(n+1)$-th level is

$$E_{n+1}^{\text{Voigt}} = (1-\varphi_n)E_n^p + \varphi_n E_n \cong \varphi_n E_n,$$

where E_n^p is the Young's modulus of the soft phase at the n-th level. When the total volume fraction of mineral $\Phi = \varphi_0 \varphi_1 \cdots \varphi_{N-1} = \prod_{n=0}^{N-1} \varphi_n$ is fixed, increasing the total number of hierarchy levels N tends to increase φ_n, allowing E_{n+1} of Eq. (9) to approach the Voigt bound E_{n+1}^{Voigt}. Therefore, larger N generally leads to higher overall stiffness of the composite.

When the staggered structure is subjected to uniaxial tension, the mineral particles are primarily under tension with protein layers in-between transfer loads primarily via shear [10]. By means of self-similar design, this feature is carried over to all hierarchical levels. Assuming that the particle aspect ratio is determined by the particle to matrix strength ratio [6,15], the tensile limiting strength at the $(n+1)$-th level depends on which phase of the n-th level fails first. If the hard particles fail first, we have $S_{n+1} = \varphi_n S_n / 2$. On the other hand, if the soft matrix fails first, in the most efficient way [30] that the stress transfer in the matrix remains uniform [6,10], we will

9

have $S_{n+1} = \varphi_n \rho_n \tau_n^p / 2$, where τ_n^p stands for the shear strength of the soft matrix. Therefore, the strength of the hard particles at the $(n+1)$-th level can be expressed as

$$S_{n+1} = \min\left(\frac{\varphi_n \rho_n \tau_n^p}{2}, \frac{\varphi_n S_n}{2}\right). \tag{10}$$

From the energy dissipation point of view, it is important that the soft matrix undergoes large deformation and sliding before the hard particles fail under tension. An optimal design is that the soft matrix should fail simultaneously with the hard particles. Under this condition,

$$\rho_n \tau_n^p = S_n, \tag{11}$$

and the tensile strength of the self-similar bone at the $(n+1)$-th level is

$$S_{n+1} = \varphi_n S_n / 2, \quad S_0 = \sigma_{th}. \tag{12}$$

For the staggered structure at the $(n+1)$-th level, the effective fracture energy should include the energy required to break both the hard particles and soft matrix of the n-th level. Therefore, it can be expressed as

$$\Gamma_{n+1} = \varphi_n \Gamma_n + (1-\varphi_n) l_n \tau_n^p \Theta_n^p, \tag{13}$$

where Θ_n^p denotes the effective strain measuring the range of deformation of the soft matrix at the n-th level. On the right-hand side of Eq. (13), the first part stands for the energy required to break the hard particles while the second part refers to the fracture energy corresponding to the soft matrix. Here the width of the fracture process zone is assumed to be on the order of the length of the hard particles l_n. For simplicity, we assume that the strain energy is primarily dissipated by the deformation of the soft matrix at any hierarchical level. Eq. (13) can be reduced to

$$\Gamma_{n+1} \cong (1-\varphi_n) l_n \tau_n^p \Theta_n^p = (1-\varphi_n) h_n S_n \Theta_n^p, \tag{14}$$

where Eq. (11) has been adopted.

Once we have calculated E_{n+1}, S_{n+1} and Γ_{n+1}, the characteristic size of flaw tolerance at the $(n+1)$-th level can be determined according to Eq. (2) as

$$h_{n+1} = \frac{\Gamma_{n+1} E_{n+1}}{S_{n+1}^2}. \tag{15}$$

Substituting Eqs. (12) and (14) into (15) yields

$$\frac{h_{n+1}}{h_n} = \frac{4(1-\varphi_n)\Theta_n^p E_{n+1}}{\varphi_n^2 S_n}, \tag{16}$$

where E_{n+1} is given by Eq. (9). From Eqs. (8) and (16), the hierarchical structure can thus be determined in the following bottom-up sequence

$$h_0 \to h_1 \to h_2 \to \cdots \to h_N = H,$$

provided that φ_n, Θ_n^p are known.

In order to demonstrate the properties of such hypothetical hierarchical material, we have performed calculations based on some specific parameter choices [6,15]. For simplicity, we assume that the volume fraction φ_n and the aspect ratio ρ_n of each level are identical, i.e., $\rho_n = \rho$, $\varphi_n = \varphi$, so that the structure becomes self-similar and the volume fraction of the hard particles in each level is related to the total fraction Φ of mineral as

$$\varphi = \Phi^{1/N}. \tag{17}$$

In addition, we assume that the soft matrix at all hierarchical level has the same elastic modulus and the same range of failure deformation, namely, $G_n^p = G_p$ and $\Theta_n^p = \Theta_p$. With these selections, the multi-level stiffness of the self-similar bone can be calculated as

$$E_{n+1} = \left[\frac{4(1-\Phi^{1/N})}{G_p \Phi^{2/N} \rho^2} + \frac{1}{\Phi^{1/N} E_n} \right]^{-1}, \quad E_0 = E_m, \tag{18}$$

while the multi-level strengths are given by

$$S_{n+1} = \Phi^{1/N} S_n / 2, \quad S_0 = \sigma_{th},$$

which leads to a simple solution

$$S_n = \Phi^{n/N} \sigma_{th} / 2^n. \tag{19}$$

Substituting Eq. (19) into Eq. (11), we have

$$\tau_n^p = S_n / \rho = \Phi^{n/N} \sigma_{th} / 2^n \rho,$$

suggesting that the shear strength of the soft matrix decreases at higher levels. Inserting Eq. (17) into Eq. (14) gives rise to the multi-level fracture energy as

$$\Gamma_{n+1} = (1-\Phi^{1/N}) h_n S_n \Theta_p, \tag{20}$$

where h_n can be determined by the flaw tolerance criterion

$$\frac{h_n}{h_{n-1}} = \frac{4(1-\Phi^{1/N})\Theta^p E_n}{\Phi^{2/N} S_{n-1}}, \quad h_0 = \frac{2\gamma E_m}{\sigma_{th}^2}. \tag{21}$$

In this fashion, all structural levels of the self-similar bone can be determined, according to Eqs. (18-21), one after another in a bottom-up sequence

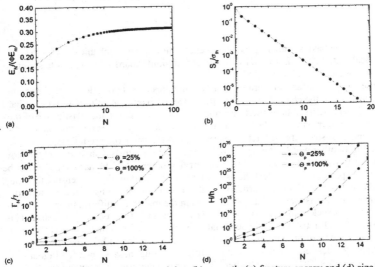

Figure 7 Variations of (a) Young's modulus, (b) strength, (c) fracture energy and (d) size of the self-similar bone with number of hierarchical levels N.

Fig. 7 shows the calculated properties of the self-similar bone as a function of the number of hierarchical levels N. In the calculations, we have assumed typical materials properties of bone as $\gamma = 1\,\text{J/m}^2$, $\Phi = 0.45$, $E_m = 100\,\text{GPa}$, $\sigma_{th} = E_m/30$, $E_m = 1000 G_p$ and consider two estimates $\Theta_p = 25\%$ and $\Theta_p = 100\%$ for the failure strain of protein. Fig. 7(a) plots the overall stiffness of the hierarchical structure normalized by the Voigt upper bound of the composite. It is seen that hierarchical design only results in a moderate increase in stiffness. After a few levels of hierarchy, the stiffness saturates at about 30% of the Voigt limit. Fig. 7(b) shows that the strength of the self-similar hierarchy drops by roughly a factor of 2 with each added level of hierarchy, decreasing by about two orders of magnitude with 6 levels of hierarchy. On the other hand, the hierarchical structures exhibit very dramatic effects on the toughness of the composite. Fig. 7(c) plots the variation of fracture energy with the number of hierarchical level for different Θ_p. It can be seen that the fracture energy increases exponentially with the increasing number of hierarchies. Fig. 7(d) plots the normalized overall size H/h_0 of the self-similar hierarchy. One can see that the flaw tolerance size of the material increases exponentially with the number of hierarchical levels. Under the selected material parameters, the flaw tolerance size of the lowest hard phase is estimated to be $h_0 = 18\,\text{nm}$. Depending upon the assumed failure strain Θ_p of protein, the flaw tolerance size increases to about 0.1 μm with only one level of hierarchy, 0.6 μm-10 μm with two levels of hierarchy, 4 μm-100 μm with 3 levels of hierarchy, 40 μm-10 mm with 4 levels of hierarchy, and 10^2-10^6 km with 9 levels of hierarchy, implying that $H \to \infty$ as $N \to \infty$. Therefore, with increasing hierarchical levels, the self-similar hierarchy can tolerate crack-like flaws of any size. These calculations demonstrate the enormous potential of a bottom-up design methodology on improving the capability of materials against crack-like flaws.

CONCLUSIONS

In this paper, we have summarized some recent studies on the mechanical principles behind hierarchical materials design via a hypothetical self-similar hard-soft composite structure mimicking the nanostructure of bone [6,15].

For the self-similar structural hierarchy, three principles need to be adopted to determine the structure at each level: the characteristic size, the aspect ratio and the volume fraction of hard particles. In the model discussed here, the first principle is to select the characteristic size of the structure to achieve uniform stress in the hard particles at failure, so as to ensure random crack-like flaws in the material/structure will not propagate until the entire structure fails near its theoretical strength. In other words, Principle I is to ensure that the stress distribution in the hard particles remain uniform so that the maximum load carrying potential of the hard material can be realized. The second principle is to select the aspect ratio of the hard particles to ensure that the shear stress in the matrix material is uniform so that the maximum load transferring capacity of the soft matrix is realized. Also, even if the shear stress in the matrix is uniform, the aspect ratio may need to be further checked to make sure that the particles do not fail before the matrix. This condition is expressed in Eq. (6). Once the potential of both hard and soft material is realized to the maximum extent, the third principle is to select the volume fraction of hard material (mineral) to achieve the desired balance between stiffness and toughness. Generally, higher mineral content promotes stiffness, while higher amount of protein promotes toughness.

12

REFERENCES

1. Currey, J.D., 1977. Mechanical properties of mother of pearl in tension. Proc. R. Soc. Lond. B 196, 443-463.
2. Currey, J.D., 1984. The Mechanical Adaptations of Bones. Princeton University Press, Princeton, NJ, pp. 24-37.
3. Jäger, I., Fratzl, P., 2000. Mineralized collagen fibrils: a mechanical model with a staggered arrangement of mineral particles. Biophys. J. 79, 1737-1746.
4. Fratzl, P., Burgert, I., Gupta, H.S., 2004a. On the role of interface polymers for the mechanics of natural polymeric composites. Phys. Chem. Chem. Phys. 6, 5575-5579.
5. Fratzl, P., Gupta, H.S., Paschalis, E.P., Roschger, P., 2004b. Structure and mechanical quality of the collagen-mineral nano-composite in bone. J. Mater. Chem. 14, 2115-2123.
6. Gao, H., 2006. Application of fracture mechanics concepts to hierarchical biomechanics of bone and bon-like materials. Int. J. Fracture138, 101-137.
7. Autumn, K., Liang, Y.A., Hsieh, S.T., Zesch, W., Chan, W.P., Kenny, T.W., Fearing, R., and Full, R.J., 2000 Adhesive force of a single gecko foot-hair. Nature 405, 681-685.
8. Autumn, K., Sitti, M., Liang, Y.A., Peattie, A.M., Hansen, W.R., Sponberg, S., Kenny, T.W., Fearing, R., Israelachvili, J.N., and Full, R.J., 2002 Evidence for van der Waals adhesion in gecko seta. Proc. Natl. Acad. Sci. USA 99, 12252-12256.
9. Yao, H., Gao, H., 2006. Mechanics of robust and releasable adhesion in biology: bottom-up designed hierarchical structures of gecko. J. Mech. Phys. Solids 54, 1120-1146.
10. Gao, H., Ji, B., Jäger, I.L., Arzt, E., Fratzl., P., 2003. Materials become insensitive to flaws at nanoscale: lessons from nature. Proc. Natl. Acad. Sci. USA 100, 5597-5600.
11. Gao, H., Chen, S., 2005. Flaw tolerance in a thin strip under tension. J. App. Mech. 72, 732-737.
12. Arzt, E., Gorb, S., Spolenak, R., 2003. From micro to nano contacts in biological attachment devices. Proc. Natl. Acad. Sci. USA 100, 10603-10606.
13. Gao, H., Wang, X., Yao, H., Gorb, S., Arzt, E., 2005. Mechanics of hierarchical adhesion structures of geckos. Mechanics of Materials 37, 275-285.
14. Gao, H., Yao, H., 2004. Shape insensitive optimal adhesion of nanoscale fibrillar structures. Proc. Natl. Acad. Sci. USA 101, 7851-7856.
15. H. Yao and H. Gao, 2007. Multi-scale cohesive laws in hierarchical materials, Int. J. Solids Struct., 44, 8177–8193.
16. A.P. Jackson, J.F.V. Vincent and R.M. Turner, 1988. The mechanical design of nacre. Proc. Roy. Soc. Lond. B 234, 415–440.
17. R. Menig, M.H. Meyers, M.A. Meyers and K.S. Vecchio, 2000. Quasi-static and dynamic mechanical response of Haliotis rufescens (abalone) shells. Acta Mat. 48, 2383–2398.
18. R. Menig, M.H.Meyers, M.A. Meyers and K.S. Vecchio, 2001. Quasi-static and dynamic mechanical response of Strombus gigas (conch) shells. Mat. Sci. Eng. A 297, 203–211.
19. W.J. Landis, 1995. The strength of a calcified tissue depends in part on the molecular structure and organization of its constituent mineral crystals in their organic matrix. Bone 16, 533–544.
20. W.J. Landis, K.J. Hodgens, M.J. Song, J. Arena, S. Kiyonaga, M. Marko, C. Owen and B.F. McEwen, 1996. Mineralization of collagen may occur on fibril surfaces: evidence from

conventional and high voltage electron microscopy and three dimensional imaging. J. Struct. Biol. 117, 24–35.

21. J.Y. Rho, L. Kuhn-Spearing and P. Zioupos, 1998. Mechanical properties and the hierarchical structure of bone. Med. Eng. & Phys. 20, 92–102.

22. S. Weiner and H.D. Wagner, 1998. The material bone: structure–mechanical function relations. Annual Review of Materials Science 28, 271–298.

23. H. Warshawsky, 1989. Organization of crystals in enamel. Anat. Rec. 224, 242–262.

24. W.Tesch, N. Eidelman, P. Roschger, F. Goldenberg, K. Klaushofer and P. Fratzl, 2001. Graded microstructure and mechanical properties of human crown dentin. Calc. Tissue Int. 69, 147–157.

25. H.D. Jiang, X.Y. Liu, C.T. Lim and C.Y. Hsu, 2005. Ordering of self-assembled nanobiominerals in correlation to mechanical properties of hard tissues. Appl. Phys. Lett. 86, 163901.

26. R.Z. Wang, Z. Suo, A.G. Evans, N. Yao, and I.A. Aksay, 2001. Deformation mechanisms in nacre. J. Mat. Res. 16, 2485–2493.

27. M.J. Buehler, S. Keten, T. Ackbarow, 2008. Theoretical and computational hierarchical nanomechanics of protein materials: Deformation and fracture. Prog. Mat. Sci. 53, 1101-1241.

28. R. Puxkandl, I. Zizak, O. Paris, J. Keckes, W. Tesch, S. Bernstorff, P. Purslow, P. Fratzl, 2001. Viscoelastic properties of collagen: synchrotron radiation investigations and structural model. Phil. Trans. Roy. Soc. Lond. B, 357, 191-197.

29. B. Ji and H. Gao, 2004. Mechanical properties of nanostructure of biological materials. J. Mech. Phys. Solids 52, 1963–1990.

30. B. Chen, P.D. Wu and H. Gao, 2009. A characteristic length for stress transfer in the nanostructure of biological composites. Comp. Sci. Tech. 69, 1160-1164.

31. H.L. Cox, 1952. The elasticity and strength of paper and other fibrous materials. Brit. J. appl. Phys. 3, 72-9.

32. J.W. Hutchinson and H.M. Jensen, 1990. Model of fiber debonding and pullout in brittle composites with friction, Mech. Mater. 9:139–163.

33. J.A. Nairn, 1997. On the use of shear-lag methods for analysis of stress transfer in unidirectional composites, Mech. Mater. 26:63–80.

34. C.L.Tucker III and E. Liang, 1999. Stiffness predictions for unidirectional short fiber composites: review and evaluation, Compos. Sci. Technol. 59:655-671.

35. M.J. Buehler, 2007. Nature designs tough collagen: Explaining the nanostructure of collagen fibrils. Proc. Natl. Acad. Sci. USA 103, 12285-12290.

36. X. Guo and H. Gao, 2005. Bio-inspired material design and optimization. IUTAM Symposium on topological design optimization of structures, machines and materials – status and perspectives, October 26–29, 2005, Rungstedgaard, Copenhagen, Denmark.

Architectured Structural Materials: A Parallel Between Nature and Engineering

John W. C. Dunlop[1], Yves J. M. Brechet[2]
[1]Department of Biomaterials, Max Planck Institute of Colloids and Interfaces, Am Mühlenberg 1, 14424, Potsdam Germany
[2]SIMAP, INP Grenoble/CNRS, 101 Rue de la Physique, BP46, 38402 St Martin d'Hères cedex, France

ABSTRACT

Nature builds materials like an architect to obtain a variety of properties with a limited number of building blocks. In contrast, engineers have access to a wide range of constituent materials to fulfil a variety of requirements. The classical degrees of freedom for controlling the properties of man-made materials are the microstructure, or the macroscopic shape. Only recently, the architecture at the millimetre scale was perceived as an efficient way of expanding the range of properties offered by bulk materials. The aim of this paper is to compare the different strategies and to outline some observations on natural materials which may serve as inspiration to develop engineering architectured materials.

Keywords:
Architectured materials, microstructure, biomimetics, design

INTRODUCTION

Facing a set of requirements, even the simplest ones (for instance the structural requirements which amount to bear a load with limited damage) is always a dilemma. Clearly one has to make the best use of matter to fulfil often conflicting requests, such as strength and damage tolerance. It is interesting to see that biological and engineering materials may represent quite different solutions of a similar problem [1].

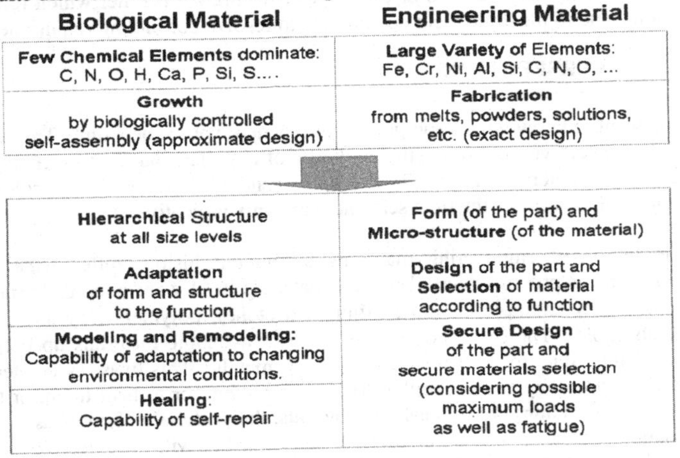

Biological Material	Engineering Material
Few Chemical Elements dominate: C, N, O, H, Ca, P, Si, S....	**Large Variety** of Elements: Fe, Cr, Ni, Al, Si, C, N, O, ...
Growth by biologically controlled self-assembly (approximate design)	**Fabrication** from melts, powders, solutions, etc. (exact design)
Hierarchical Structure at all size levels	**Form** (of the part) and **Micro-structure** (of the material)
Adaptation of form and structure to the function	**Design** of the part and **Selection** of material according to function
Modeling and Remodeling: Capability of adaptation to changing environmental conditions.	**Secure Design** of the part and secure materials selection (considering possible maximum loads as well as fatigue)
Healing: Capability of self-repair	

Figure 1: A parallel between biological and engineering materials. Reproduced from [1, 2]

While the engineer has access to the whole set of elements in the Mendeleev table and with all their variety of chemical bonds (metallic, ionic or covalent), Nature has had to develop relatively low temperature processes, and therefore is limited either to organic chemistry, or to ionic solutions. As a result, only ceramics and polymers are used as major constituents for structural materials in living bodies. The different strategies are also reflected in the processes by which materials are made [1, 2]. Nature "grows" its structural materials in an approximate design, while the engineer "shapes" the structural material according to a predefined form. This difference also reflects a different treatment of the information necessary to build a component or an organ: Nature follows a "rule" (the genetic code) while the Engineer follows a very stiff set of requirements (the design). Last but not least, the strategy to deal with damage is also very different. The engineer typically tries to dimension the components so that the damage during the expected lifetime is as limited as possible, while nature has developed a whole range of adaptive and healing processes. Not surprisingly, these very different limiting conditions have led to different strategies in meeting the mechanical requirements with an optimal use of matter. For example, one major strategy observed in natural materials is to structure the material over many length scales [2-6].

When facing a demanding situation from a structural viewpoint such as ensuring a sufficient strength with a minimum weight, the engineer typically has a more reduced choice. The engineer can either play with the microstructure below the micron scale, for instance by inducing a precipitation inside an alloy, or by designing the component shape on the other end of the size scale for instance in creating I beams, sandwiches etc. The variety of elements that can be used has opened the route to "alloy design", in defining the microstructure, and the variety of processes has allowed more and more efficient shapes. Only relatively recently, with the emergence of fiber reinforced composites [7], or foamed materials [8], the architecture at the intermediate scale was recognized as a possible strategy to meet such requirements [9]. And not surprisingly, the limitations on the use of these architectures, comes from the difficulties in processing them.

Of course these classifications over simplify reality. In the first place, structural components have often multiple functions in nature. Bone, in addition to acting as a structural member, is a reserve for calcium for instance [10] so that the "optimization" concept has to be manipulated with care. Beside, engineers are progressively discovering the virtues of architecturing at a millimetric scale (foams, felts, ...) and even try to develop healing strategies [11, 12]. The main purpose of this paper is to draw a parallel which reflects these recent evolutions, and to illustrate them on attempts to resolve conflicting requirements.

A HISTORICAL PERSPECTIVE

"If you want to know where you go, consider where you come from". This Senegalese proverb is a good incentive to examine the evolution of materials, on "historical" times scales [13] for engineering materials (Figure 2) and more generally speaking on "geological time scales for biological materials (the time scale here are only indicative).

The evolution of engineering materials throughout history reflects the ability of various civilizations to process them. The first materials used were ones that were already existing in nature: for example ceramics (flint, and stone), polymers (leather, tendon and wood) or metals (gold). The expansion of the range of available materials up until the 20th century was mainly due to a better handling of metals. This was associated with a development of new energy sources, either to elaborate them, or to heat them, or to deform them. Only in the 20th century a new class of materials, the artificial polymers associated with the oil civilisation, appeared in a significant amount, together with technical ceramics made possible by high temperature / high pressure processes. It is worth noticing that the development of foams, essentially polymer foams, has populated the region of low density materials very efficiently. It is also interesting to observe that, before the foams appeared, the range of density spanned by engineering materials just reflects the available range of elements to make them.

The situation is quite different for biological materials. The building blocks of natural materials, consisting mainly of low density peptides and sugars with some minerals, have been around for a long time. The main tool nature has to evolve new materials, is to modify the ways in which these components are put together, through chemistry, and microstructure, but also through architectural design at intermediate length scales. This is commonly seen in many biological systems that use hierarchical structuring to achieve optimal performance at all length scales (2). The predominance of atoms of low atomic number (C, H, O, N, Ca, Si and P) means the high density range of materials is never densely populated. Many natural materials also are low density cellular materials, which allows organisms to optimise materials usage also allowing for fluid and cell transport [14]. The real "inventors" in biological materials were the arthropods, which developed three main classes of biological materials: the polymer / ceramic composites, the cellular materials, and the natural elastomers. Cellular materials developed very much in structural component of fishes and in plants. Insects "invented" a variety of "high tech" natural fibres such as silk, and the mammals, very conservative, simply developed leather on their own.

The very concept of architecture is relatively recent in engineering. At the macroscopic level, in spite of the intuitions of the cathedral architects, the major developments date from the 19th century where metallic alloys became the major industrial structural material. At the mesoscopic level, foams and fibre reinforced polymers were born thanks to petrochemistry, developed in the 20th century. The scope for developing architectured engineering materials relying both on the variety of constitutive materials (which is much larger than in biology) and on the variety of architectures (exemplified by natural materials but which presents serious processing difficulties for the engineer), is simply enormous.

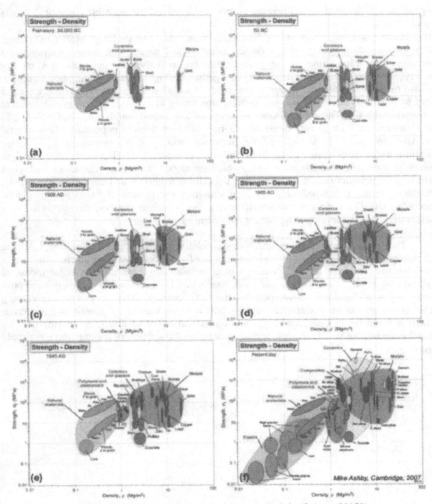

Figure 2: Evolution of engineering materials (from ref [13])

STATIC LOADING: SOLVING CONTRADICTIONS

It is always tempting to devise new strategies to expand the range of available materials, and joining the variety of "building blocks" available to the engineer with the variety of natural architecture has a certain Faustian flavour. However, the engineer is also concerned with the practical advantages of such an extension: "understand in order to realize" is his motto. There is a very good reason to explore the possibilities of architectured engineering materials, which is to "fill the gaps" in the "materials properties charts" [15]. Most properties of materials are controlled i) by the nature of the chemical bond ii) by the packing density of atoms iii) by the defects in a crystalline structure. These are rather limited controlling parameters and as a consequence, there are a number of properties which are intrinsically contradictory. For instance the strength and the elastic moduli both depend on the chemical bond: having a material which is both strong and flexible is not easy. Toughness is related to the elastic modulus and to the dissipated energy during crack propagation, therefore a high toughness tends to be incompatible with a large elastic limit. Wear resistance is related to hardness, and therefore tends to be contradictory with toughness. In the following we consider successively these three pairs of conflicting requirements, and the solutions proposed both in engineering and in biology.

Keeping Strong while Controlling Flexibility

One engineering strategy to develop structures with high strength and high flexibility is to play on the geometry at the component level, and more specifically to "prefragment" the material. This enables the intrinsic strength of the material to be kept in tensile loading, and allows flexibility in bending (e.g. Figure 3). In contrast the external geometry can also be designed for minimum flexibility as seen in the classical examples of I-beams and tubes, which have an optimal bending stiffness and strength. Design at the component level is of course also used by Nature and can be seen in well-known structures such as leaves, stems and stalks etc [16].

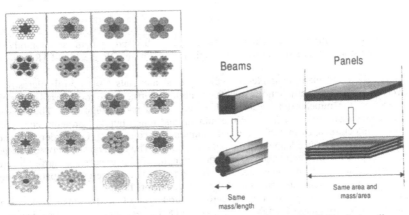

Figure 3: Engineering strategies to combine strength in tension and flexibility in bending

Tough and Strong

The contradiction between strength and toughness is a difficult problem for the engineer, which has been addressed for example by the steel manufacturers principally through control of the microstructure and chemistry. The intrinsic contradiction is illustrated in Figure 4, which shows a clear "anti correlation" between formability and strength. Martensitic steels, very strong, have a very poor formability, while IF steels, very formable, have a very low strength. The contradiction comes both from the inability for very strong materials to blunt cracks formed at defects by plasticity, but also from the unstable macroscopic plastic behaviour. When the flow stress is very large, the work hardening rate is unable to stabilize the tendency for flow localization, which is the physical meaning of Considère criterion. By modifying the microstructure, engineers have managed to provide an extra work-hardening through a "composite effect" - the strategy proposed by dual phase steels. Even cleverer is the solution of the TRIP steel, which uses plasticity induced phase transformations to provide an increasing fraction of non deformable phase. This then leads to a work hardening behaviour which adapts itself to the current flow stress. Microstructural design therefore helps resolve the contradiction of formability versus strength and allows for new regions of materials space to become accessible.

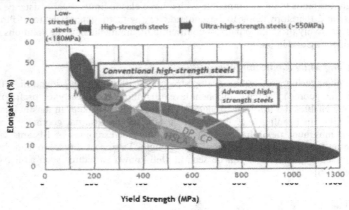

Figure 4: The intrinsic contradiction between formability and strength for automotive steels. The strategies to bypass this contradiction: Dual Phase (DP) steels and TRIP steels

There are many examples in which natural materials also control microstructure to overcome intrinsic contradictions, such as combining stiffness and toughness [2], for example as seen in the mineralised tissues, such as bone, and nacre (Figure 5). These tissues are composites made up of stiff but brittle mineral and tough but soft proteinaceous polymers. The architecture allowing this unusual combination consists at first order of nano-sized mineral platelets embedded within the polymer matrix. The load transfer between the polymer and the platelet is obtained by shear, the strength of the platelets is large due to their nanometric size [17], and the elongation which permits a large energy dissipation (toughness) is provided by the shear deformation of the collagen layers [18].

Both nature and the engineer use microstructural control at the small scale to address materials contradictions of toughness/strength and toughness/stiffness. The engineer optimises phase distributions for the DP and TRIP steels, and nature the mineral platelet/collagen structure for bone. Clearly both strategies, based on microstructure development, are possible only through some plasticity in the material which permits kinematic composite hardening to develop.

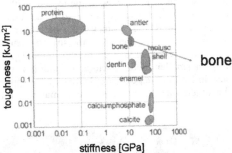

Figure 5: The architecture allowing a good compromise between toughness and stiffness in bone from [2].

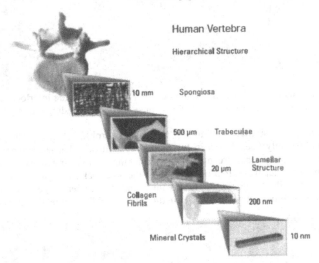

Figure 6: Bone as an example of hierarchical structuring in a natural material [19].

We have seen in the previous examples that both nature and the engineer optimise properties of a component by modifying the microstructure as well as the macrostructure. In addition nature also optimises properties by controlling design on intermediate scales, something that has recently sparked much interest in the design of engineering materials such as truss structures and foams [8, 20]. This additional design on the intermediate level is well illustrated by the hierarchical structure of bone (Figure 6). The mineralised collagen fibrils (at the nanometre scale), form lamellar structures on the micron scale, which in turn make up the material of trabeculae on the millimetre scale [2]. This structuring, allows for multiple toughening mechanisms in bone: crack-bridging, crack-deflection, and microcracking to name a few examples [5]. Although the hierarchical structure of bone is one of the most studied [3, 5, 21-23], the principle it illustrates of using energy dissipation at all length scales to increase toughness seem to be rather general [2, 6, 24, 25].

But what if we are considering a ceramic, which is intrinsically stiff, brittle and shows no plasticity whatsoever? If the engineer wants to design for very high temperature conditions, ceramics become more or less unavoidable. A possibility to make such structure "damage

21

tolerant" has been recently proposed in [26, 27] with the topologically interlocked structures (Figure 8). In these structures the material is architectured on an intermediate level, by pre-fragmented the ceramic into blocks. These geometrically lock each other in place, due to externally applied compressive stresses. Any crack in the structure can only propagate within a single block. Damage tolerance is ensured both by this limited propagation, and by the reduced probability to have a smaller block fracture. In addition, such a material can be "tunable" since its stiffness in bending depends on the externally applied compressive stress as well as on the detailed geometry and specific constitutive materials for each block [26, 27].

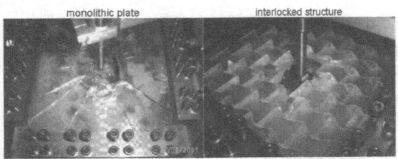

Figure 8: Interlocked materials : comparison between an osteomorphic interlocked Plexiglass structure and the equivalent monolithic plate loaded in bending, after [26].

Interestingly, a similar system to the osteomorphic blocks with interlocking architectures has recently been found in the natural system of a Turtle carapace (Figure 9) [28]. The shell is a bony structure made of interlocking ribs separated by a thin layer of protein. It is thought that the soft protein layer in the interlocking region between the ribs gives flexibility to allow the reptile to breathe, and yet under large deformations the interlocking sections block, stiffening the structure and providing protection against external aggression [28].

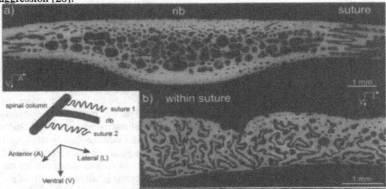

Figure 9 : a natural example of interlocked structure : the red ear turtle carapace from [28]

Wear resistant and tough

The last example we will briefly consider is wear resistance. The traditional engineering solution to wear resistance is by surface hardening, by cementation and nitriding in steels for

22

example. The price to pay for that is a tendency to crack, which causes major problems, together with fatigue, for ball bearings. Nature has developed a clever solution to this problem of dealing with for teeth. Teeth are made up of two main tissues, hard wear-resistant enamel on the outside which sits on more compliant dentine. The two materials are in contact at the Dentine-Enamel Junction (DEJ), although the details of the material in this layer are not well known its function is to transfer loading from the hard enamel layer to the softer dentine [29]. In addition it is thought this layer also functions to prevent crack propagation by blunting the cracks of the brittle layer. A similar strategy could be envisaged for ball bearings, having a soft layer under the cemented one, for instance via a decarburation performed before cementation.

DYNAMIC LOADING: ADAPTATION

The examples listed above illustrate the development of architectures to fulfil conflicting requirements for a prescribed loading. There is another situation for which nature is especially efficient: when the loading evolves during the life of the organism due to either growth or a changing environment. This is the case for instance for wood. During the lifetime of a tree the mechanical requirements of new tissue changes as branches grow thicker due to the addition of annual growth rings. Trees can respond to changes in loading by increasing stiffness through modified cellulose orientations in the new tissue [2, 30]. Bones have an alternative strategy to deal with changing mechanical environments with the (re)modelling cycle as described by the so-called Wolff-Roux Law [31, 32]. Bone cells (osteoblasts) produce tissue where it is needed and remove (osteoclasts) tissue where it is not. This process allows bones to adapt to increased loading and is responsible for loss in bone in low-gravity environments. The adaptability of wood and bone is a specific faculty of biological system which supposes both sensors for the load, and actuators which direct matter transport. The details of the mechanisms are not known in most cases, but the ability of adaptative growth is a key feature of natural systems. In contrast, engineering systems, in spite of some attempts under the name of "intelligent materials" or more recently "self-healing materials", are not very good at that. At best they can have sensors embodied which signal the damage, or even, in some very specific cases, ability to heal the damage before it becomes critical [11, 12]. But up to now, a fully self healing engineering system is still to be developed. Cartilage provides a wear resistance coating for about 70years in human beings (before arthritis comes as a reminder that nothing last for ever) but this is possible not because of the materials properties, but because of its ability to regenerate. The teeth provides a counter example , since there is no active self healing in it, but the contact conditions are by far less demanding too. Certainly the key difference between biological architectured materials and engineered architectured materials relies in the current limitation with imbedded repair mechanisms, and this is a route for further development.

FUTURE STRATEGIES FOR ENGINEERS

In this brief overview, we have outlined a parallel between biological and engineering materials. Central to the efficiency of biological materials is a control of architecture on many hierarchical levels, but also the ability of nature to grow, adapt and heal. In contrast, engineers have used for decades architecture at the macroscopic level, and microstructure design at the microscopic level to obtain efficient solutions to design problems. With the need to use matter and energy in a more efficient way, the sets of requirements for the engineers has become more and more demanding, and often contradictory properties are searched for. It is in this context of "filling the gap" in the "materials properties space" that architectured multimaterials have recieved a recent impulse. The different contributions of this symposium give a glance into a very wide range of possible applications. Although Nature can provide inspiration for the development of this new class of materials, a key issue is how to process them. Combining different materials classes, tailoring appropriate geometry is already possible. Hierarchical architecture is, up to now, rather limited (at least compared to the situation in natural materials) but is not impossible. For instance multifilament conductors used in high strength magnets are architectured on at least 5 levels of structures by multiple successive wire drawings [33]. Adaptative growth and self healing strategies are the main challenges for which it is not clear that a viable engineering solution is possible, except in very specific and emblematic cases.

ACKNOWLEDGMENTS

The authors would like to thank P. Fratzl, M.F.Ashby, J.D.Embury and J.Estrin for enlightening discussions.

REFERENCES

[1] P. Fratzl, Journal of the Royal Society Interface 4 (2007) 637-642.
[2] P. Fratzl, R. Weinkamer, Progress in Materials Science 52 (2007) 1263-1334.
[3] K. J. Koester, J. W. Ager, R. O. Ritchie, Nature Materials 7 (2008) 672-677.
[4] R. K. Nalla, J. H. Kinney, R. O. Ritchie, Nature Materials 2 (2003) 164-168.
[5] H. Peterlik, P. Roschger, K. Klaushofer, P. Fratzl, Nature Materials 5 (2005) 52-55.
[6] J. Aizenberg, J. C. Weaver, M. S. Thanawala, V. C. Sundar, D. E. Morse, P. Fratzl, Science 309 (2005) 275-278.
[7] T.-W. Chou, Microstructural design of fiber composites, Cambridge University Press, Cambridge, 1992.
[8] L. J. Gibson, M. Ashby, Cellular Solids: Structure and Properties, Cambridge University Press, Cambridge, 1997.
[9] M. Ashby, Y. Bréchet, Acta Materialia 51 (2003) 5801-5821.
[10] A. M. Parfitt, Bone 10 (1989) 87-88.
[11] S. R. White, N. R. Sottos, P. H. Geubelle, J. S. Moore, M. R. Kessler, S. R. Sriram, E. N. Brown, S. Viswanathan, Nature 409 (2001) 794-797.
[12] K. S. Toohey, N. R. Sottos, J. A. Lewis, J. S. Moore, S. R. White, Nature Materials 6 (2007) 581-585.
[13] M. Ashby, Philosophical Magazine Letters 88 (2008) 749 - 755.
[14] L. J. Gibson, Journal of Biomechanics 38 (2005) 377-399.
[15] M. Ashby, Philosophical Magazine 85 (2005) 3235-3257.
[16] U. G. K. Wegst, M. F. Ashby, Journal of Materials Science 42 (2007) 9005-9014.
[17] H. J. Gao, B. H. Ji, I. L. Jager, E. Arzt, P. Fratzl, Proceedings of the National Academy of Sciences of the United States of America 100 (2003) 5597-5600.
[18] H. S. Gupta, J. Seto, W. Wagermaier, P. Zaslansky, P. Boesecke, P. Fratzl, Proceedings of the National Academy of Sciences of the United States of America 103 (2006) 17741-17746.
[19] P. Fratzl, Hierarchical structure and mechanical adaptation of biological materials, in: Learning from Nature How to Design New Implantable Biomaterials, Kluwer Academic Publishers, 2004, pp. 15-34.
[20] A. G. Evans, J. W. Hutchinson, N. A. Fleck, M. F. Ashby, H. N. G. Wadley, Progress in Materials Science 46 (2001) 309-327.
[21] K. Tai, M. Dao, S. Suresh, A. Palazoglu, C. Ortiz, Nature Materials 6 (2007) 454-462.
[22] S. Nikolov, D. Raabe, Biophysical Journal 94 (2008) 4220-4232.
[23] P. Fratzl, Nature Materials 7 (2008) 610-612.

[24] J. Keckes, I. Burgert, K. Frühmann, M. Müller, K. Kölln, M. Hamilton, M. Burghammer, S. V. Roth, S. Stanzl-Tschegg, P. Fratzl, Nature Materials 2 (2003) 810-813.

[25] D. Raabe, C. Sachs, P. Romano, Acta Materialia 53 (2005) 4281-4292.

[26] A. V. Dyskin, Y. Estrin, E. Pasternak, H. C. Khor, A. J. Kanel-Belov, Acta Astronautica 57 (2005) 10-21.

[27] Y. Estrin, A. Dyskin, E. Pasternak, S. Schaare, S. Stanchits, A. Kanel-Belov, Scripta Materialia 50 (2004) 291-294.

[28] S. Krauss, E. Monsonego-Ornan, E. Zelzer, P. Fratzl, R. Shahar, Advanced Materials 21 (2009) 407-+.

[29] P. Zaslansky, A. A. Friesem, S. Weiner, Journal of Structural Biology 153 (2006) 188-199.

[30] A. Reiterer, H. Lichtenegger, S. Tshegg, P. Fratzl, Philosophical Magazine A 79 (1999) 2173-2184.

[31] J. Wolff, Das Gesetz der Transformation der Knochen. Translated as: The Law of Bone Remodelling, Springer Verlag, Berlin, Germany, 1892.

[32] W. Roux, Die züchtende Kampf der Teile, oder die 'Tielauslese' im Organismus (Theorie der 'funktionellen Anpassung). Wilhelm Engelman, Leipzig, Germany, 1881.

[33] K. Han, A. C. Lawson, J. T. Wood, D. Embury, R. B. Von Dreele, J. W. Richardson, Philosophical Magazine 84 (2004) 2579-2593.

Cellular and Fibrous Materials

Mater. Res. Soc. Symp. Proc. Vol. 1188 © 2009 Materials Research Society   1188-LL02-03

The Effect of Cellular Architecture on the Ductility and Strength of Metal Foams

K. R. Mangipudi and P. R. Onck

Zernike Institute for Advanced Materials, Applied Physics Department, University of Groningen, Nijenborgh 4, 9737AG, Groningen, The Netherlands

ABSTRACT

A multiscale finite element model has been developed to study the fracture behaviour of two-dimensional random Voronoi structures. The influence of materials parameters and cellular architecture on the damage initiation and accumulation has been analyzed. The effect of the solid material's strain hardening, relative density and architectural randomness on the ductility and fracture strength of the cellular solid are investigated. The results suggest materials-design directions in which the heat treatment, the solid material properties, its microstructure and the cellular architecture can be tuned for an optimized performance of cellular materials.

INTRODUCTION

In many engineering materials for structural applications the ultimate material would be the one that is both strong and tough/ductile. The traditional approach in metallurgy is to employ alloying and/or heat treatment to modify the physical microstructure of the metal (grain-size, impurities, second phase particles, etc.). Compared to dense materials, cellular materials have a two-level microstructure: (i) the solid microstructure of the constituting material and (ii) the cellular architecture. The latter describes the architectural information on how the solid material is distributed into space, forming a three-dimensional interconnected network. The cellular architecture puts in additional degrees of freedom which opens the exciting opportunity for a materials-by-design approach for cellular metals.

The overall fracture behaviour of metal foams is closely related to the microstructure of the cell-wall material. In open-cell foams made by investment casting, grain-boundary-covering, plate-like precipitates were found to be the primary cause for a knock-down in ductility. Ductility-enhancing heat treatments often result in reduction of the yield stress and an increase in hardening capacity. The overall behaviour of the structure depends not only on the underlying solid material microstructure, but also sensitively on the cellular architecture, e.g. the cell size and shape distribution, the cross-sectional geometry of the strut, the strut connectivity and its relative density. The goal of this work is study these dependencies using a multiscale modelling framework that takes all these ingredients into account and enables optimal design of cellular materials.

MODEL

A finite element model based on a Voronoi description of the cellular structure is used. Layered Euler-Bernoulli beam elements in a co-rotational framework [1] are used to model the strut deformation. Figure 1 outlines the modeling approach and the exchange of information among various length scales. Each beam element contains fibers or integration points at which the axial strain and curvature increments from the beam finite element, together with a stress-strain law for the solid, will be used to update the stress state across the thickness of the rectangular strut. Based on the nonlinear stress state, the tangent modulus at the integration points is known from the material stress-strain behaviour. The elements of the symmetric stiffness matrix of the beam element are obtained by properly integrating the tangent modulus across the thickness [1]. The softening during failure at the material point is given by a stress-displacement relationship which ensures that always a fixed amount of fracture energy per unit area of cross section of the strut Γ_0 is dissipated. Loading-unloading conditions are incorporated at these integration points by keeping track of the history of the stress evolution at the material point. The elastic-plastic material properties, such as the Young's modulus E, yield stress σ_Y and the power law strain hardening exponent N_s and the specific fracture energy Γ_0 can be obtained from a tension test of the bulk material or a single strut. Failure of fibers in compression is allowed to achieve mesh independency.

Random Voronoi structures are generated with uniform strut thickness for every relative density. A Young's modulus $E = 70$ GPa, yield stress $\sigma_Y = 41$ MPa and failure strain $\varepsilon_F = 8\%$ are used in all the simulations. Converged results were found for 40 elements per strut, 20 fibers per element for 16x20 cells (width x height) Voronoi structures.

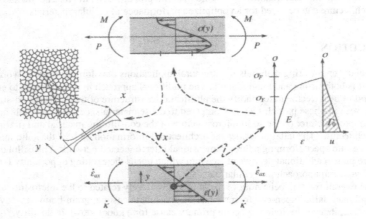

Figure 1: Multiscale modeling approach: communication between different length scales.

RESULTS

Damage initiation and accumulation

A typical overall stress-strain curve is shown in Fig. 2. After an initial elastic deformation, yielding of struts near the strut junctions (triple points) occurs in which the outer fibers reach the yield stress (denoted by point A on the stress-strain curve). Since most of the structure (strut volume) is still elastic, the overall modulus of the structure remains close to the elastic modulus. With increasing applied strain, more struts yield at triple points and yielding inside the struts spreads across the strut thickness. When the cell wall material can strain harden, stresses in the neighborhood of the triple points increase, causing yielding along the strut length developing an extended plastic zone. For high strain hardening materials, this plastic zone may cover the entire length of some struts depending on relative density and loading of the strut. With most of the triple points yielded, the stress-strain curve enters the nonlinear regime which increases the strains within the structure and strains are rather distributed for structures under uniaxial tension (see Fig. 2). However, it has been observed in incremental strain maps (not shown here) that there is a preferential growth of strains in certain locations. At any instant, the strains in these regions are much larger than the current mean strain over the structure. These regions will become the prospective locations for the damage.

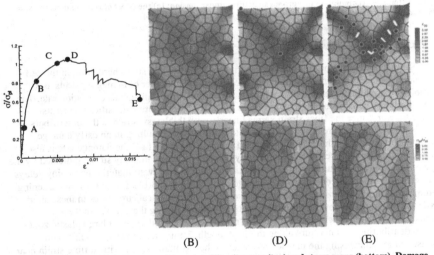

Figure 2: Typical stress-strain curve and associated Strain maps (top) and stress maps (bottom). Damage locations are identified with circles and failed (removed) struts are marked in white on the strain map.

When the fiber stress in a strut reaches the critical fracture stress of the cell wall material, damage is initiated (point C on the stress-strain curve) in one of the large strain accumulated regions (see Fig. 2). However, the structure loads with further strain as other parts of the structure can still carry the load. Soon few more struts reach the failure stress in these high strain regions forming a fracture band. As a result, unloading due to damaging struts takes over the loading due to elastically and/or plastically deforming struts and eventually a peak is formed in the stress-strain curve (at point D). During the progressive damage, stresses in the structure get redistributed (see the stress maps in Figure 2) to other undamaged parts of the structure and may lead to damage initiation in those regions. After the peak, the overall stress decreases monotonically and strain accumulation is localized within the fracture band(s). Meanwhile, the damage process is completed in some struts and starts to form a crack as can be seen in Fig. 2.

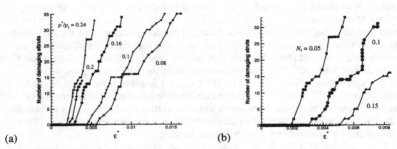

(a)   (b)

Figure 3: Damage accumulation in Voronoi structures under tension: (a) the effect of relative density (N_s = 0.05) and (b) the effect of material strain hardening (relative density = 0.24).

Relative density and strain hardening

In this section we present the effect of relative density and strain hardening in the cell wall material on the fracture behaviour. Figure 3 shows the number of damaging struts as a function of the applied overall strain, which is a measure of the damage accumulation rate. The effect of relative density can be seen in Figure 3(a). At higher relative densities, the struts become thicker. Since the fiber strain varies linearly with thickness, strain at the outer fibers exceeds the critical fracture strain at low bending strains, thus resulting in an early damage initiation. This is clearly the case in Figure 3(a). Due to the same fact, the damage rate is also enhanced at higher relative densities with early damage initiation in many struts, indicated by the slope of the curves in Figure 3(a). In contrast, increasing the cell wall material hardening delays the damage initiation and decreases the damage rate (see Figure 3(b)). For low strain hardening solid material, the plastic zone is small and resembles a hinge. Local curvatures in these short plastic zones can easily develop to initiate damage without raising the stresses in their neighborhood. On the other hand, for a high strain hardening solid material large plastic zones develop distributing the curvature along the strut length. Strains near the triple point cannot increase without increasing the plastic zone size and hence attaining the critical fiber strain near the triple point requires greater overall strains.

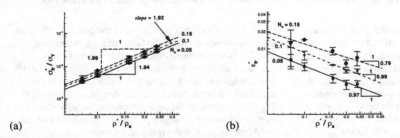

(a)   (b)

Figure 4: The effect of density and solid material hardening on (a) peak stress and (b) peak strain.

Figure 4 shows scaling of the peak stress and peak strain with relative density for different strain hardening solid materials. Similar to the plastic collapse stress, the peak stress also shows a strong power law scaling with the relative density. The exponent of the power law scaling is found to lie between 1.92 and 1.99 (comparable to the power of 2 for the plastic collapse stress). It can also be observed from Figure 4(a) that increasing the strain hardening exponent also increases the peak stress. The peak strain is also enhanced by increased strain hardening of the solid material as shown in Figure 4(b). However, it decreases with increasing relative density and exhibits an inverse power law scaling with an exponent between -0.99 and -0.79.

Randomness of the cellular architecture

The effect of randomness on elastic properties has been reported in the literature [2, 3]. Both the uniaxial and hydrostatic yield stresses of a two-dimensional cellular structure are reported to decrease with increasing randomness [2], while the Young's and shear moduli are reported to increase with randomness [3]. In this section, we discuss the effect of randomness on the fracture properties for different relative densities and for fixed solid material properties. Figure 5 shows the behaviour of peak stress and peak strain as function of randomness parameter δ. The randomness parameter δ has been defined [3] such that δ = 0 for a fully random Γ-Voronoi and δ = 1 for a regular hexagonal structure. Unlike the plastic collapse stress reported in the literature, the peak stress does not show any appreciable variation with the randomness. On the other hand, the peaks strain exhibits a clear increase with decreasing randomness (i.e., with increasing δ) and is pronounced at low relative densities.

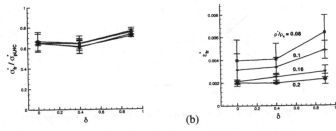

(a) (b)

Figure 5: The effect of randomness on (a) the peak stress and (b) the peak strain.

DISCUSSION

Materials for structural engineering applications require a good combination of ductility and strength. Conventional metallurgical techniques, aimed at enhancing the strength, involve alloying and heat treatments, for example tempering of Al-alloys. To enhance the strength, the primary goal of these techniques is to produce second phase particles and grain boundary precipitates which create obstacles for the movement of dislocations. Unfortunately, at the same time, these precipitates and grain boundary precipitates knockdown the ductility in foams while increasing the strength [4, 5]. A high ductility is desirable in designing the structural components to undergo sufficient plastic deformation before catastrophic failure. As shown in the previous section, the relative density also increases the strength of a cellular structure at the expense of ductility. Increasing the strain hardening of the solid material can be very attractive to gain both strength and ductility at the same time. However, the gain in strength by enhancing the solid's strain hardening is small compared to enhancing the relative density, due to its strong scaling. However, it provides another degree of freedom to counteract the disadvantage of loss of ductility through an increased relative density. Aluminium alloys in tempered (T6) conditions possess a low strain hardening capacity besides loosing ductility in the process of increasing the yield strength. Increasing the relative density will result in a greater loss of the ductility as the negative slope of the peak strain increases with decreasing N_s in Figure 4(b). Hence, this heat treatment will produce a severe knockdown in the overall ductility. We have found in a separate study that increasing the yield strength of the solid material also increases the overall strain hardening capacity of the structure [6]. The effect of brittle precipitates had also been shown to be deleterious in terms of both ductility and strength [4, 5]. Apart from the solid material properties, the cellular architecture also plays an important role in determining the overall mechanical response. As shown in the previous section, decreasing the structural randomness increases the ductility while relatively unaffecting the peak stress. Hence, the knowledge of the effects of various material and structural parameters allows for the possibility of an optimal design of cellular materials for a given structural application. These observations are summarized pictorially in Figure 6.

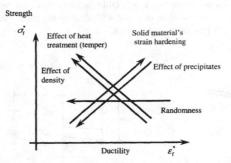

Figure 6: Summary of the effect of various parameters on ductility and strength of a cellular solid.

REFERENCES

1. M.A. Crisfield, "*Non-linear finite element analysis of solids and structures: Vol-1*", John-Wiley & Sons, 1991.
2. C. Chen, T.J. Lu, N.A. Fleck, *J. Mech. Phys. Solids* **47**, 2235(1999).
3. H.X. Zhu, J.R. Hobdell, A.H. Windle, *J. Mech. Phys. Solids* **49**, 857(2001).
4. E. Amsterdam, P.R. Onck, J.Th.M. De Hosson, *J. Mat. Sci.* **40**, 5813(2005).
5. K.R. Mangipudi, E. Amsterdam, J.Th.M. De Hosson, P.R. Onck, "Porous metals and metallic foams", *Proceedings of MetFoam-2007*, ed. L.P. Lefbvre et. al., (DEStech Publications Inc., 2007), pp363.
6. K.R. Mangipudi, S.W. van Buuren, P.R. Onck, "The microstructural origin of the strain hardening in open-cell metal foams", *to be submitted for publication*.

Mater. Res. Soc. Symp. Proc. Vol. 1188 © 2009 Materials Research Society   1188-LL06-05

Modelling the mechanical and thermal properties of cellular materials from the knowledge of their architecture

Maire E.[1], Caty O.[1,2], Bouchet R.[2], Loretz M.[3] and Adrien J.[1]

[1] Université de Lyon, INSA-Lyon, MATEIS CNRS UMR5510 F-69621 Villeurbanne, France
[2] ONERA Châtillon, 92322 CHATILLON, France.
[3] Université de Lyon, INSA-Lyon, CETHIL Villeurbanne, France.

ABSTRACT

This paper shows different examples where the architecture of cellular materials has been determined exactly using 3D X ray computed tomography. The images were then subsequently used to generate FE meshes reproducing the architecture as exactly as possible. The FE meshes where in turn used to simulate the mechanical (monotonous and fatigue compression) and the thermal (radiative properties) behavior of the studied materials.

INTRODUCTION

Playing games with the architecture of cellular materials opens thousands of different possible solutions for obtaining various properties. This large amount of solutions offered with a single composition by the arrangement of the solid and gaseous phase is extremely rich. The design engineers can start to imagine different combination of properties for a same given material. This important amount of possible solution can however possibly become a drawback: it is not humanly feasible to analyze experimentally all the different possibilities. In this context, modeling the properties of cellular materials from the knowledge of their architecture allows the different solutions to be screened. The more promising solutions can subsequently be analyzed experimentally and their properties verified.

The problem is more and more tricky today because the properties required for certain applications are multi-functional. In a wide variety of applications, there is a need for the prediction of the mechanical properties but also and simultaneously of the thermal, acoustical, electrical, chemical properties... etc. All these properties are strongly affected by the structure. The present paper will show how one can determine with a common frame the thermal and mechanical properties of architectured cellular materials. This approach is based on the exact knowledge of the microstructure achieved using 3D images of the materials. X-ray tomography has appeared recently to be a very powerful tool allowing to characterize the architecture of structural materials [1, 2] or with a lower resolution, cellular materials [3-7]. The low X-ray absorption of these materials, composed of a large amount of air, permits large specimens to be studied using the technique. The second advantage of this method is to allow local and global large deformations to be imaged non destructively so the important buckling, bending or fracture events appearing during the eventual mechanical deformation can be visualized.

The architecture has a strong effect on the behavior but the modeling methods proposed so far do not fully capture this effect. The seminal work by Gibson and Ashby [8] has permitted to underline that the relative density is a key parameter but the scatter in the properties of cellular solids can not simply be explained using this approach based on cells with an idealized morphology. The tomography technique allows to capture architectural differences between materials exhibiting identical relative densities (cell size distributions, cell wall defects...). This suggests to use this new kind of information as an input for models which should describe better the peculiarities of the behavior of these materials.

This paper will show how the 3D images of cellular materials can be quantitatively used to extract key parameters for the architecture. Different possible methods for using tomography results as inputs for models, especially Finite Element meshes, will be discussed. The meshes obtained will be used in mechanical and thermal problems. The mechanical problems are the prediction of the static properties of an ERG aluminum foam, but also of the fatigue properties of metal hollow sphere structures. The second example is the simulation of the thermal properties of the same ERG aluminum foam.

EXPERIMENTS AND MATERIALS

The images used in the present study were obtained using a standard laboratory tomograph available in the MATEIS laboratory at INSA Lyon. It is a commercial product sold by the Phoenix X-ray company. It is built inside a lead self protected cabin. The dimensions of the lead cabin are 1.2x1.5x2 m approximately. It contains two principal components. The X-ray source is an open transmission nanofocus X ray tube operated between 40 and 160 kV. The X ray detector is a Paxscan™ amorphous silicon flat panel initially developed for medical application. It is composed of 1920 rows and 1536 lines of sensitive pixels, the size of each is 127 x 127 μm². The minimum voxel size (of the order of 0,6 μm) is obtained when the sample is the closest possible to the source. The spatial resolution is altered by the finite focus size through the well known "geometrical blur" effect. In practice, the minimum resolution ever obtained in a tomographic scan using our setup is about 1.5 μm. Fig. 1 shows the architecture of the samples used in the two examples developed in this paper.

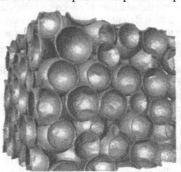 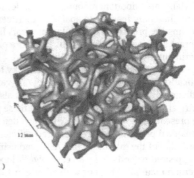

Fig. 1a : 3D X ray reconstruction of the Metal Hollow Sphere Structure used in the mechanical example. Resolution 26μm.

Fig. 1b : 3D X ray reconstruction of the ERG Al foam used in the thermal example. Resolution 20 μm.

FINITE ELEMENT MESHING

It can be seen from Fig 1 that the architecture of structural materials is most of the time rather complex. The appropriate tool which allow to perform the calculation in such a complex situation is the FE method. The basis of the approach explained in the present paper is to produce FE meshes picturing the 3D images of the architecture as exactly as possible. We wish firstly to discuss here and to present our own investigations on the different methods which can be used to generate FE meshes based on tomography images of the actual three dimensional architecture of the studied materials.

Three different strategies can be used to produce meshes picturing the 3D data sets. The most straightforward technique consists of replacing each voxel of the 3D data set by a cubic element of the same size [9,10]. A second technique, developed for cellular materials [11] and more recently used for a bulk material [12] consists of meshing the outer surface of each phase in the microstructure using triangular elements. The solid part of the microstructure defined by the outer surfaces is meshed in this method by an advanced front technique [11] using tetrahedral volume elements starting from a surface mesh. If the material is a cellular material, that is, contains a large amount of porosity, another method can be used provided that a graph (or skeleton) of the solid phase can be calculated. Instead of completely meshing the solid phase, it is sometimes sufficient for the structures of these materials to be simplified by means of beam [13,14] or shell elements. Fig. 1b actually shows for example the surface mesh describing exactly the geometry of the foam used for the thermal example in the present paper. It will be discussed later that this surface mesh is sufficient to calculate the radiative properties of any cellular material. The tetrahedral mesh that can be generated from this surface mesh to discretized the solid part of the material can in turn be used to simulate the conduction behavior of the material of interest.

When the thickness of the solid phase is very small compared to the size of the cells, it is probably more efficient to think about an appropriate way of generating shell elements picturing the architecture of the cellular material (method 3). The method presented in [15] is based on the skeletonization of the solid phase by thinning [16]. The final goal is to determine, by image processing, the median plane of each wall. The process described works with binary images and with closed cells. Each cell of the gaseous phase has first to be labeled. Each label can then be dilated isotropically through the solid phase until it encounters the neighboring dilating label. At the end of this dilation procedure, the solid phase has been eliminated and the frontier between two neighboring labels lies exactly in the middle of the previously existing wall between these two cells This frontier can be meshed by a marching cube algorithm using surface (shell) triangular elements and can then be simplified to reduce the number of triangles while preserving a good description of the surface [17]. The linear triangular elements produced are then firstly imported into the FE code used [18]. They are then translated into quadratic elements using an available subset of the FE software (translation from S3 to STRI65 in Abaqus™). The measure of the thickness was done in the present work using a mathematical morphology procedure called "3D Granulometry" (described in [19]) involving erosion/dilation steps of the binary image. The structural element used for the operation is a sphere in our case. The treatment provides a map of the wall thickness over the entire specimen. The second step calculates the coordinates of the barycentre of each element in the mesh and attributes the measured thickness read in parallel on the map. The relative density of the mesh must sometimes be slightly corrected by a coefficient applied uniformly to the thickness of all the elements in order to be in exact agreement with the overall solid fraction in the material.

MECHANICAL SIMULATION

ERG foam meshed with tetrahedral elements

Provided that a 3D mesh of the solid phase of the ERG foam can be generated, it is then straightforward to both calculate the global mechanical behavior of the foam and then to analyze the local stress field inside this complex microstructure. Fig. 2 shows the three stress strain curves calculated from the 3D images by loading numerically the sample in three different directions. The plateau stress reached in these simulations is consistent with the usual strength of these materials experimentally measured in the literature (around 0.8 MPa). Note also that the simulation shows that the foam is anisotropic. This sort of measurement can of course not be achieved experimentally on a same piece of material because the first compression would destroy the architecture of the sample. These simulations can then be used to estimate the global behavior of the samples. It becomes very easy to calculate the effect of the mechanical behavior (modulus, yield strength, strain hardening) of the solid phase on the global resulting properties of the cellular material. The mechanical loading in difficult experimental conditions such as multiaxial can also be rather easily simulated. These calculations are however probably more interesting for accessing the local values of the stress or strain field inside the architecture for a given mechanical loading. This is exemplified in the next subsection.

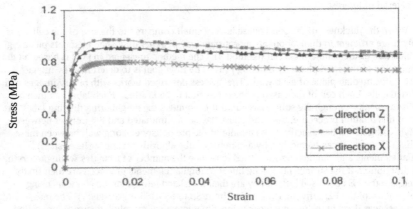

Fig. 2 : global stress strain behavior calculated for the ERG material in three perpendicular directions

MHSS meshed with shell elements for fatigue properties

When shell elements are used, the number of elements is much smaller than in the previous case for a same architecture. In the example, a very large tomographic bloc has been calculated – dimensions of the sample are 575*615*650 voxels with a resolution of 26µm. The behavior of the bulk material is modeled in the simulation by an elastoplastic law. We chose to use adapted coefficients for the wall porosity rate: E = 150 GPa and Poisson ratio ν = 0.3. E is estimated by multiplying the steel modulus with the volume fraction of metal (about 75 %). The plastic behavior of the solid phase is discretized and modeled by a proof stress vs. plastic strain table chosen to reach a good fit for the global behavior of the foam while keeping the value in the range commonly given for porous sintered stainless steel i.e. {(140 MPa , 0); (153.4MPa, 0.035);(187Mpa, 0.1); (210Mpa, 0.2)} [20]. Fig. 3 shows the deformed contour plot of the Von Mises stress in one of the slices of the structure calculated using the procedure described above applied to the MHSS sample shown in Fig. 1 in compression--compression fatigue. The computer used all through the present study is a standard personal computer (IntelR XeonTM, CPU 3.2 GHz, RAM 2Go). It is interesting to note that the new meshing procedure presented here allowed us to calculate with this modest computer the total volume of the sample tested experimentally. As a consequence, the effects of the free surfaces of the sample are completely accounted for.

A local validation of the shell model regarding the fatigue local deformation modes can be obtained by the comparison with the actual deformation kinematics of the material observed by X-Ray micro tomography. Fig. 3 also shows the comparison of the experimentally deformed state in the corresponding slice. The elastic displacements have been magnified to be more visible. The mesh has been obtained using the XRCT scan of the sample at the initial state. For the sake of simplicity, the calculation has been performed using just one loading cycle, the behavior of the walls being assumed to be elastic. Despite the differences between the experimental conditions and the hypothesis used to perform the calculation, one can note two interesting things validating the proposed method at the local scale:

- the local deformation patterns predicted by the model and observed in the deformed slice are very close (the slight differences are due to the fact that it is always a bit difficult to find the same slice in the initial and in the deformed 3D image);
- the cracks are observed to appear experimentally in regions where the stresses are high in the elastic calculation (darker regions in the contour plot).

Finally, the resulting calculation can be post processed to determine the weak points of the microstructure under a given (an arbitrary) load. Our final goal is to obtain a fatigue criterion identified via this post processing of the local stress field on one of our sample and to check blindly the validity of this criterion for all the other samples. This part of the work is still in progress

MODELLING OF THE THERMAL PROPERTIES

The modeling of the conductive part of the thermal properties of a cellular materials can be performed exactly in the same way as described above for the mechanical properties. Calculation of the heat conduction using a FE mesh is a straightforward problem where one should carefully account for the proper boundary conditions. The only tricky part is simply to generate the proper mesh. Coquard et al. [21] have for instance used voxel based meshes to conduct this type of calculation. We will then rather focus in this section on the modeling of the radiative properties. This can be achieved by a Monte Carlo type approach where light rays are randomly traced inside an exact description of the outer surface of the solid phase such as this depicted in Fig 1b. Fig. 4 schematically shows how these different rays can be absorbed, reflected or directly transmitted.

From the counting of the number of rays of each kind and using the optical properties of the solid matrix, it becomes possible to access the radiative properties, which are the extinction coefficient β, the scattering albedo ω and the scattering phase function Φ, appearing in the radiative transfer equation. Table 1 shows extinction coefficient obtained for the ERG Al foam for a great number of rays, typically 200000 in this case. This coefficient is calculated by the average of the results obtained by the previous method for wavelengths ranging from 5 µm to 15 µm (for a radiative heat transfer calculation at 300 K). The experimental radiative properties obtained by spectrometric measurements are also represented. The agreement between the experimental extinction coefficient and that resulting when accounting for the tomographic structure of the foam is good.

Material	$\beta_{tomography}$ (m−1)	$SD_{tomography}$ (%)	$\beta_{experimental}$ (m−1)
Al-foam	172	6	170

Table 1: Comparison between the extinction coefficients obtained from tomography and experimental measurement, with the standard deviation.

According to the standard deviation, the extinction coefficient is almost independent on the wavelength. This confirms that geometrical optic laws (and thus Monte Carlo method) can be applied to determine the radiative characteristics of the foam. In addition, we can study the influence of the mesh on the results. Table 2 shows the relative errors compared with fine mesh, obtained for β calculated using coarser meshes. We can note that the errors on the extinction coefficient by using medium and coarse meshes are below 2% and 5% respectively. If these errors remain low, the mesh step is however important and it must be verified that the simplification of the mesh structure complies with the real one.

Mesh	Intermediate	shapeless
Error vs fine mesh (%)	0.8	2.7

Table 2: Influence of the mesh size on the error during the calculation of β.

CONCLUSIONS

Different ways exist to produce FE meshes picturing exactly the actual architecture of cellular materials. These FE meshes can then be used for different sorts of simulation. This paper has shown examples of a same ERG foam for which both mechanical and thermal properties have been calculated using these meshes. The case of MHSS for which dedicated meshes have to be produce to allow the calculation of larger samples has also been presented. These calculations are now more and more reliable. This method is a valuable to optimize the multi-functional properties of these material. It is also a useful tool for helping the designers to tailor the architecture in order to obtain the desired properties for a given application.

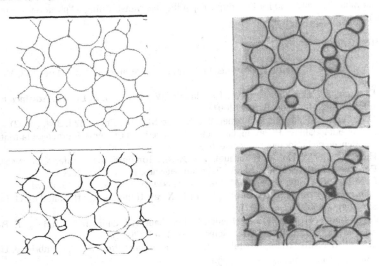

Fig 3a : FE simulation of the compression loading of the MHSS using the mesh generated from the images with shell elements of controlled thickness. Contour plot of the von Mises stress (high values of the stresses appear darker)

Fig 3b : Experimental comparison of the initial (top) and fatigue deformed state (bottom) of the same slice as in figure a using XRCT. The width of the bloc is 15 mm.

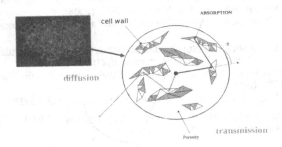

Fig. 4 : schematic diagram showing the principle of the Monte Carlo type modeling of the radiative properties.

41

ACKNOWLEDGMENTS

The authors which to thank Yves Brechet for valuable scientific discussion in the domain of cellular materials, and also for his unique capability for transforming a "pseudo volunteer" process into an actual commitment.

REFERENCES

1. J.Y. Buffière, E. Maire, P. Cloetens, G. Lormand, R. Fougères, Acta Mater., 47, 1613 (1999).
2. E. Maire, J.Y. Buffière, L. Salvo, J.J. Blandin, W. Ludwig, J.M. Létang, Advanced Engineering Materials, 3, 539 (2001).
3. O.B. Olurin, M. Arnold, C. Körner, R.F. Singer, Mat. Sc. Eng, A328, 334, (2002).
4. A. Elmoutaouakkil , F. Peyrin, G. Fuchs, ,"Proceedings of 7 ème Journée de la Matière Condensée de la Société Française de Physique", 645, (2000).
5. H.P. Degisher, A. Kottar, F. Foroughi, In "X-Ray Tomography in Material Science", ed. Baruchel, Buffière, Maire, Merle, Peix, Ed. Hermes, Paris, 165, (2000).
6. R. Müller, T. Bösch, D. Jarak, M. Stauber, A. Nazarian, M. Tantillo, S. Boyd ,"Proceedings of the SPIE, Developments in X-ray Tomography III, edited by U. Bonse, San Diego, 4503, 189 , (2001).
7. J.A. Elliott, A.H. Windle, J.R. Hobdel, G. Eeckhaut, R.J. Oldman, W. Ludwig, E. Boller, P. Cloetens, J. Baruchel , J. Mat. Science. 37, 1547, (2002).
8. L.J. Gibson, M.F. Ashby, "Cellular Solids: Structure and Properties", Cambridge, UK: Cambridge Univ. Press, 2nd ed. (1997).
9. A.P. Roberts, E.J. Garboczi, *Acta Mater.* 49, 189 (2001).
10. D. Ulrich, B. Van Rietbergen, H. Weinans, P. Ruegsegger, *J. Biomech.* 31, 1187 (1998).
11. S. Youssef, E. Maire, R. Gaertner, Acta Mater. 53, 719 (2005).
12. K. Madi, S. Forest, M. Boussuge, S. Gailliègue, E. Lataste, J.-Y. Buffière, D. Bernard, D. Jeulin, Comput. Mater. Sci. 39, 224 (2007).
13. J.A. Elliott, A.H. Windle, J.R. Hobdel, G. Eeckhaut, R.J. Oldman, W. Ludwig, E. Boller, P. Cloetens, J. Baruchel, J. Mater. Sci. 37, 1547 (2002).
14. P.R. Onck, R. van Merkerk, J.T.M. De Hosson, I. Schmidt, Adv. Eng. Mater. 6, 6 (2004).
15. O. Caty, E. Maire, R. Bouchet. Acta Mater. 56, 5524, (2008).
16. Z.Liang, M.A.Ionnidis, I.Chatzis, *Journal of Colloïd and Interface Sciences*, 221, 13, (2000).
17. http://www.tgs.com/support/amira_doc/Documentation/amira41UsersGuide.pdf
18. ABAQUS, *Finite element program theory manual*, (1998)
19. E. Maire, P. Colombo, J. Adrien, L. Babout, L. Biasetto, Journal of the European Ceramic Society 27, 1973, (2007).
20. W-S. Lee, C-F. Lin, T-J. Liu, *Material Characterization.* 58, 363, (2007).
21. R. Coquard, M. Loretz, D. Baillis. Conductive Heat Transfer in Metallic/Ceramic Open-Cell Foams. Adv engng mater. 10, 323, (2008).

Mater. Res. Soc. Symp. Proc. Vol. 1188 © 2009 Materials Research Society   1188-LL02-04

A Preliminary Study on Cell Wall Architecture of Titanium Foams

Nihan Tuncer[*], Luc Salvo[+], Eric Maire[†] and Gursoy Arslan[*]

*Anadolu University, Dept. of Mater. Sci. & Eng., Iki Eylul Campus, 26555, Eskisehir, Turkey

[†] INSA Lyon MATEIS lab, Bat St Exupery 25 Av Capelle, 69621, Villeurbanne cedex, France

[+]INPG Grenoble, Laboratoire SIMaP/GPM2, BP 46 38402 Saint Martin d'Heres, France

ABSTRACT

Bio-inspired architectures, especially metallic foams, have been receiving an increasing interest for the last 10 years due to their unusual mechanical properties. Among commonly dealt foamed metals, like aluminum and steel, titanium possesses a distinctive place because of its high strength-to-weight ratio, excellent corrosion resistance and biocompatibility. In this study, Ti foams were produced by a very simple and common method, sintering under inert atmosphere with fugitive space holder. Removal of the space holder was conducted by dissolution in hot deionized water which makes it possible to minimize contamination of Ti. Sintering of remaining Ti skeleton at 1300 °C offered a wide range of properties and cost savings. The effects of the processing parameters such as sintering temperature and powder characteristics on the 3D foam architecture were investigated by using X-ray microtomography (μ-CT). Use of bimodal Ti powders caused a decrease in final theoretical density when compared to the ones prepared with the same amount of space holder but with monomodal Ti powders. It was also observed that the use of bimodal Ti powders decreased compressive strength at a fixed relative density, by introducing pores into the cell walls.

INTRODUCTION

Although the vast majority of research on the processing and characterisation of metallic foams has focused on melt processing of Al and steel foams [1-4], Ti and its alloys are very attractive since they combine unique properties such as high strength-to-weight ratio, excellent corrosion resistance and biocompatibility. These properties offer special application areas including biomedical applications and aircraft and spacecraft engineering [5].

As in all branches of material science, it is important to link physical properties of cellular solids to their density and complex microstructure, in order to understand how such properties can be optimized for a given application. In literature, there exist simple predictions of mechanical properties which are mainly based on the porosity amount and on an assumption concerning the architecture of the solid phase [6]. However, very often, certain amounts of discrepancy exist between the predicted and experienced results which arise from underestimation of real architectural affects. In studies carried out so far, certain tendencies in compressive strength and energy absorption with regard to Ti foams were determined [7]. However, it is important to explain the reason of the deviation from ideality and what happens if parameters other than porosity, such as the internal architecture, are further changed.

As stated in the recent literature [8], statistical models which allow variation of the density and structure seem to be ideally suited especially for random cellular materials. Therefore, direct imaging of the foams with different structures followed by quantitative measurement of structural parameters constitutes the first step of property prediction by statistical modeling. The term *structural parameter* refers to the elements that, in combination, form the internal architecture of a cellular structure. Cell wall porosity, cell size distribution, cell wall thickness distribution, tortuosity, cell sphericity, cell coordination, size of interconnections and total surface area are among the structural parameters to be measured in order to comment on mechanical properties [9-10].

Powder metallurgical route, by nature, allows tailoring internal architecture in a very wide range. However, this high degree of freedom brings along a very high number of processing parameters, such as sintering temperature, metallic powder size and distribution, spacer size and morphology, which affect structural parameters simultaneously. Therefore the processing parameters should be watched carefully to address property alteration correctly.

In this study, the influence of porosity location on compressive strength was investigated. Sintering efficiency was interfered by adding coarser Ti powders into the metallic power-spacer mixture in order to alter cell wall porosity. Structural changes were investigated by μ-CT.

EXPERIMENT

High melting temperature and extreme affinity with atmospheric gases render it difficult to process Ti in liquid form whereas powder metallurgy provides easier processing conditions. This method also allows fine tuning of the internal architecture. Ti foams were manufactured by space holder method using Grade 2 commercially pure (cp) Ti powders, supplied by Alfa Aesar, Germany, (Table 1). Angular shaped Ti powders were preferred in order to increase the packing efficiency without using a binder, since preforms pressed from spherical powders tend to collapse during removal of the space holder. The particle size range of the carbamide $(CO(NH_2)_2)$ space holder was arranged to vary between 0.3 - 0.5 mm by sieving (Fig. 1).

Figure 1. SEM images of (a) carbamide particles (b) fine Ti powders and (c) coarse Ti powders.

Table 1. Particle size and specific surface area of the Ti powders used.

	Fine Ti	Coarse Ti
d_{10} (μm)	13,5	95,2
d_{50} (μm)	27,5	145,1
d_{90} (μm)	50,4	221,7
Specific Surface Area (m²/g)	0,28	0,12

70 vol. % space holder and balance Ti powders were dry mixed in a ZrO_2 medium using ball milling for 1 hour. The mixtures were first pressed at 100 MPa uniaxially in steel dies to

ease handling. The preforms were then cold isostatically pressed at 300 MPa in order to prevent collapse of the green body during space holder removal in water.

Carbamide was leached out completely, as confirmed by weighing the specimens before and after the space holder removal step, in hot deionized water without destroying the green skeleton which then were sintered in a tubular furnace at 1300 °C for 2 hours under a flowing argon (Ar) gas atmosphere after being vacuumed and purged with Ar gas three times.

Compression test with quasistatic (10^{-3} s^{-1}) strain-rate was conducted on samples having cross-sectional areas of around 55 mm^2 and length-to-width ratios of 1.5. Compression tests were carried out at room temperature with an Instron 5581 model Universal Testing Machine. Strains were calculated from the cross-sectional displacement of the crosshead. Compression cages were lubricated to minimize friction between the sample and the cages. Compression strengths were determined using 0.2 % offset method. The average of 4 measurements was taken as the mean.

μ-CT has been used because it is one of the most efficient direct imaging techniques. The studied material was cut into 3 mm x 3 mm cross-sections and scanned with 3 μm resolution. Scanning of the samples was carried out with a standard tomography equipment located at the MATEIS laboratory at INSA, Lyon. The tomograph was operated at 90 keV and 150 μA. The radiographs obtained after 3D x-ray scanning were reconstructed and 400 x 400 x 400 voxel representative volumes were selected for each sample and were characterized quantitatively by the use of Image J and Aphelion softwares in SIMAP laboratories at INPG, Grenoble. To be able to compute structural parameters on individual cells, the cells were separated using a 3D watershed algorithm.

RESULTS and DISCUSSION

The decrease in total powder surface area which resulted from the addition of coarser Ti particles, caused a lower final density and a lower percent shrinkage in all directions which implies lowered sintering efficiency on cell walls (Table 2). In the table, h, w and t refer to height, width and thickness of Ti foam samples, respectively. The average of 4 measurements was taken as the mean. It was also observed that addition of a third powder type, besides carbamide and fine Ti powders, diminishes homogeneous mixing which, in turn, resulted in a higher standard deviation in final density.

Table 2. Green and final densities and shrinkage percentages of the foams.

	Density after spacer removal (%)	Final Density (%)	Shrinkage after sintering (%)		
			h	w	t
No coarse Ti addition	27,15	38.45 ± 0.45	10,14	9,96	10,30
25 % coarse Ti addition	27,37	36.11 ± 1.40	8,48	8,62	9,00
50 % coarse Ti addition	28,08	35.55 ± 1.25	7,39	7,59	6,70

The coarser Ti addition was observed to cause a strength and homogeneity drop (Table 3). The strength drop with the addition of 50 % coarse Ti powders was around 50 %. Higher standard deviation in density and compressive strength values which was caused by improper mixing when coarser powders were added was also observed. Since the strength decrease may also have resulted from density decrease, mechanical tests were repeated with the foams having the same total porosity but different cell wall porosity. It was observed that porous cell walls resulted in a lower compressive strength for the same porosity (Fig. 2). However the strength drop was determined as only 17 %. This means that there are other factors, which may have aroused from internal architecture alteration, affecting mechanical properties besides porosity.

Table 3. Density and compressive strengths of the foams.

	Final density (%)	Compressive Strength (MPa)
No coarse Ti addition	38.45 ± 0.45	60.11 ± 5.24
25 % coarse Ti addition	36.11 ± 1.40	46.80 ± 10.35
50 % coarse Ti addition	35.55 ± 1.25	32.55 ± 10.26

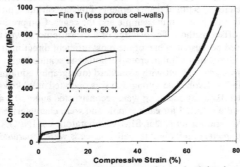

Figure 2. Compressive strength-strain plots of the two foams both having 60 % porosity.

In order to investigate these changes in structure, μ-CT images were used. Cell size distribution, cell sphericity, cell wall porosity and cell wall thickness distribution were calculated by the use of image analysis.

The tomography slices (Fig. 3) revealed that coarse Ti addition changed the internal structure as well. Increasing cell wall porosity was also obvious even by visual observation. Besides becoming more porous, cell walls seemed to have a wider thickness distribution. Without the addition, cells maintained the spacer morphology whereas a roughening on cell faces and deterioration in the internal structure were observed with the addition.

Volumetric and number frequencies of cell size calculations were carried out by taking the diameter of the equivalent sphere displaying the same volume as cell size (Fig. 4). Number frequencies were calculated in order to have an idea about the cell wall porosity which was very high in number when compared to the spacer cells. Cells having a diameter below 50 μm were assumed to be the cell wall porosity since the spacer size was above 300 μm. Number distribution showed that both the number and size of cell wall porosity increase with the addition of coarser Ti. Volumetric distribution, on the other hand, was plotted to observe the cells created by the spacers. While no difference in mean cell size of no coarse Ti addition and 25 % coarse Ti addition foams was observed, further decrease in sintering efficiency on cell walls provided by 50 % coarse Ti addition caused an increase on mean cell size. Physical explanation of this cell enlargement may be that the presence of pores on the cell phases, besides within the cells, increased the individual cell volume. Another effect which takes attention is that the foams produced with coarser Ti addition had a shoulder on the volumetric cell size distribution whereas the ones produced with only fine Ti showed a monomodal distribution. This was probably because of the poorer packing efficiency when coarser Ti was added.

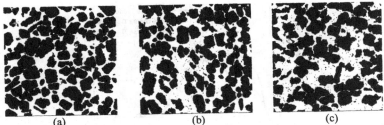

Figure 3. Tomography slices of the foams with (a) no coarse Ti addition, (b) 25 % coarse Ti addition and (c) 50 % coarse Ti addition.

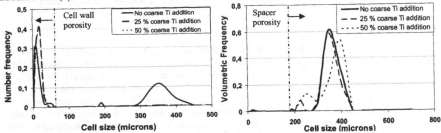

Figure 4. (a) Number, (b) volumetric cell size distributions of the foams produced.

Volume percentage of cell walls occupied by pores showed that both number and vol.% of cell wall pores increased due to poor sintering efficiency (described as the % increase in relative density due to sintering) when coarse Ti powders were added to the batch (Table 4). Calculations showed that pore sphericity decreased with coarse Ti addition, in agreement with the visual observation on the tomography slices (Fig. 3). It was already reported that wavy imperfections, cell face corrugations, cell shape variations decrease the strength and stiffness of the foams [11]. Increased roughness of pore faces might act as stress concentrators, which in turn, may have contributed to the strength drop.

Cell wall thickness measurements performed using 3D granulometry showed higher numbers of scatter in distribution and a higher mean cell wall thickness with coarser Ti addition but there was no pronounced tendency with the increase of addition (Fig. 5). Although an additional image pretreatment was applied to ensure the removal of cell wall porosity which is misleading for this computation, the measurements will be repeated with more samples in order to render the results clearer.

Table 4. Cell wall porosity and sphericity values of the foams produced.

Coarse Ti addition	No of pores detected within the cell walls	% cell wall volume occupied by pores	Sphericity (pores larger than 200 µm)	Sintering efficiency (%)
No	386	0,015	0.68	27
25 %	2921	0,148	0.67	24
50 %	3148	0,364	0.60	20

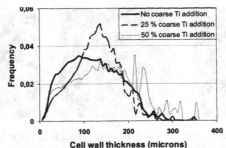

Figure 5. Cell wall thickness distributions of the foams with and without coarse Ti addition.

CONCLUSIONS

The influence of porosity location on compressive strength was investigated on Ti foams produced via powder metallurgical route, which offers a high degree of freedom in tailoring the foam architecture. Sintering efficiency was interfered by adding coarser Ti powders into the metallic powder-spacer mixture in order to alter cell wall porosity. Structural changes were investigated by μ-CT. Cell wall porosity, created by the addition of coarse Ti powders, caused a strength drop in the samples having the same amount of total porosity. However, decreased cell sphericity and cell face roughening with addition of coarser Ti powders played an important role in strength decrease. Investigations showed that when employing powder metallurgy, the processing parameters should be watched carefully to address property alteration correctly.

ACKNOWLEDGMENTS

The authors would like to thank the Scientific Research Commission of Anadolu Uni. for financing this work under the contract no. 040231 and Jerome Adrien from INSA Lyon Mateis.

REFERENCES

1. Banhart, J., Progress Mater. Sci., **46**, 559 (2001).
2. D. K. Balch, J. G. O'Dawyer, G. R. Davis, C. M. Cady, G. T. Gray, D. C. Dunand, Mater. Sci. Eng. A 391 (2005) p. 408.
3. C. Park, S. R. Nutt, Mater. Sci. Eng. A **323**, 358 (2002).
4. Jiang, B., Zhao, N.Q., Shi, C.S., Li, J.J., *Scripta Mater*, **53**, 781, (2005).
5. Salimon, A., Bréchet, Y., Ashby, M. F., Greer, A. L., J. of Mater. Sci., **40**, 5793 (2005).
6. Gibson, L. J. Ve Ashby, M.F., *Cellular Solids*, Second Edition, Cambridge Solid State Science Series, ISBN **0-521-49560-1**, (1997).
7. N.Tuncer, G.Arslan, J. of Mater. Sci. **44**, 6, 1477 (2009).
8. Roberts, A.P., Garboczi, E.J., J. Mech. Physics of Solids, **50**, 33 (2002).
9. L. Salvo, P. Cloetens, E. Maire, S. Zabler, J.J. Blandin ,J.Y. Buffiere, W. Ludwig, E. Boller, D. Bellet, C. Josserond, Nuclear Instruments and Methods in Physics Res B **200**, 273, (2003)
10. Ariane Marmottant, PhD Thesis, INPG, December 8th, France, (2006).
11. A.E. Simone, L.J. Gibson, Acta Mater. **46**, 3929 (1998).

Mater. Res. Soc. Symp. Proc. Vol. 1188 © 2009 Materials Research Society   1188-LL02-02

Fibers Networks as a New Type of Core Material: Processing and Mechanical Properties

Laurent Mezeix[1, 2], Christophe Bouvet[2], Serge Crézé[3], Dominique Poquillon[1]
[1] CIRIMAT, Université de Toulouse, INPT-ENSIACET, 118 route de Narbonne, 31077 Toulouse, France
[2] LGMT, IGM, Université de Toulouse, LGMT, bât 3PN, 118 route de Narbonne, 31062, Toulouse, France
[3] DMSM, Université de Toulouse, ISAE campus SUPAERO, 10 avenue Edouard Belin BP 54032, 31055, Toulouse, France

ABSTRACT

Fibers networks materials have been elaborated from different fibers: metallic fibers, glass fibers and carbon fibers. Cross-links have been achieved using epoxy spraying. The scope of this paper is to analyze the mechanical behavior of these materials and to compare it with available models. The first part of this paper deals with elaboration of fibers network materials. In the second part, compression tests are performed. The specific mechanical behavior obtained is discussed.

INTRODUCTION

Entangled materials can be made from natural materials (wool, cotton…) as well as artificial ones (steel wool, glass wool…). Bonded metal fibers network materials offer advantages [1-7] for use like heat exchanger [8] or insulation [9]. Indeed, they present a low relative density, high porosity and simplicity of production by cost-effective routes with considerable versatility concerning metal composition and network architecture.

On the other hand, sandwich panel consists of two thin skins separated by a thick core. Core material is usually in the form of honeycomb, foam or balsa. Recently, a novel type of sandwich has been developed with bonded metallic fibers as core material [10-16]. This material presents an attractive combination of properties like high specific stiffness, good damping capacity and energy absorption. Metal fibers are bonded with a polymeric adhesive [15] or fabricated in a mat-like form and consolidated by solid state sintering [12,16].

Entangled cross-linked carbon fibers have been also studied for use as core material by Laurent Mezeix [17]. Indeed entangled cross-linked carbon fibers present many advantages for application as core material: open porosity, possibility to mix different fibers to optimize properties or possibility to reeve electric or control cables on core material. Only a few studies are devoted to the mechanical behavior of material made from entangled cross-linked fibers [16, 18].

In the present paper, fibers have been chosen in function of their nature, applications and costs. Due to their high performance, carbon fibers are intensely used in many applications: aeronautics, sport equipments or high performance vehicles. Because of their low cost, glass fibers are widely used in many applications like thermal insulation for pipe, building. Bonded metal fiber network are used as core material [10-16]. The purpose of the present study is to analyze the mechanical behavior and to model the initial stiffness made from three types of cross-linked fibers.

EXPERIMENT

Material elaboration

Carbon fibers (200 Tex) consist in a yarn of stranded carbon filaments. Fibers were provided by Toho Tenax. The Young modulus of the carbon fiber is 240 GPa. Fibers diameter is 7 µm and the initial epoxy coating represents 1 wt%. Assuming than the coating is uniform, then, its thickness is about 30 nm. Stainless steel fibers were provided by UGITECH. Fibers diameter is 12 µm, the Young modulus is 180 GPa. Glass fibers are obtained from yarns (600 Tex) that were provided by the company PPG Fiber glass. Fibers diameter is 12 µm and their Young modulus is about 73 GPa.

For aeronautical applications, many sandwich-panel skins are made using carbon/epoxy prepreg. That the reason why epoxy resin has been chosen for cross-linking fibers. Epoxy resin was provided by the company SICOMIN. The provided resin has a low viscosity (285 mPa.s) and polymerization duration is 2 hours at 80°C. For all the tests carried out during this work, specimens are carefully weighted using SARTORIUS balance (±10µg). Resin is heated up to 35 °C to decrease viscosity and thus allows a better vaporization.

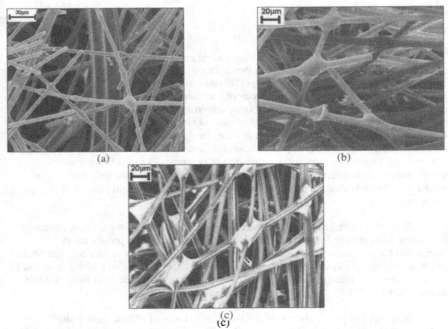

(a)   (b)

(c)

Figure 1. SEM observation on materials made with (a) cross-linked carbon fibers, (b) cross-linked glass fibers and (c) cross-linked stainless steel fibers.

A paint spray gun (Fiac UK air compressors) is used to spray epoxy. Materials were observed using a Scanning Electron Microscope (LEO435VP) operating at 15 kV. Fibers are cut so the fibers length equals 40 mm. For core application, one key parameter is the density. A previous study has shown that the yarns size needs to be decreased by separating filaments to obtain the lowest final density possible [19]. In the present work, separation of yarns was obtained thanks to a blower room. Then, the entangled fibers are cross-linked. In that case, epoxy is sprayed using paint spray gun during the final minutes of the entanglement. We have chosen to perform tests on fiber networks materials with fibers density of 150 kg/m³. As the volume of the mould is known and the fibers mass is carefully weighted, the volumetric density is controlled. It is important to notice as the density of bulk carbon in the carbon fibers (1760 kg/m³) is lower than the glass fibers one (2530 kg/m³), which is lower than the stainless steel one (7860 kg/m³), the relative volumetric density of the tests materials differs. Figure 1 shows typical SEM observations on networks made of carbon, glass and stainless steel fibers. We can notice clearly joints between filaments.

<u>**Compression tests**</u>

The quasi-static compressive response of entangled fibers is measured using a screw-driven test machine MTS with a 5 kN load cell. The sample is introduced between the punches and the compression test is then carried out (no lateral confinement). The sample size is 30x30x30 mm³. The punch velocity is $v_0 = 1.8$ mm.min⁻¹ corresponding to a nominal strain rate of $\dot{\varepsilon} = 10^{-3}$ s⁻¹. To analyze the experimental results, we used the usual definition for the true strain and true stress.

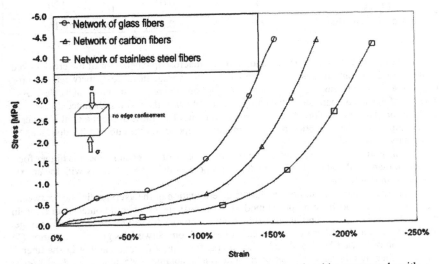

Figure 2. Comparison of compression stress/strain curves of architectures made with entangled fibers cross-linked with epoxy resin (initial density = 150 kg/m³).

Cross-linked fibers samples have been tested in compression. Density used for each type of fibers is 150 kg/m^3. The initial fiber density of the material is determined as explained above. The volume fraction of fiber differs between the three types of materials. 150 kg/m^3 correspond to a volume fraction of 1.9% for the material made of stainless steel fibers, to 5.9% for the material made of glass fibers and to 8.5% for the material made of carbon fibers. Resin is sprayed during entanglement, which is weighted again after epoxy polymerization. So the additional mass of resin can be measured and corresponds to 50 kg/m^3.

Figure 2 shows comparison during compression tests between materials made with cross-linked carbon fibers, stainless steel fibers and glass fibers. The better stiffness is achieved with the glass fibers. Furthermore, the plateau level [20] of stress during compression is quite high. Densification is the dominant mechanism for strain above 70%. The initial stiffness of the networks glass fibers is 8.4 MPa, 1.1 MPa for carbon fibers and 0.6 MPa for stainless steel fibers. The average stress level of the plateau before densification is 0.8 MPa, 0.4 MPa and 0.25 MPa for respectively glass, carbon and stainless steel fibers.

Table 1: Comparison of distances between cross-links.

Material	$<c>$, number of contacts per fiber	Average distance between joints : calculated value $d_{av} = L/<c>$ [mm]	Average distance between joints observed by SEM [mm]
Network of stainless steel fibers	127	0.31	0.30
Network of carbon fibers	974	0.04	0.20
Network of glass fibers	395	0.10	0.15

In order to get a better understanding of the macroscopic behavior of the cross-linked architecture, important microscopic information is the average distance d_{av} between two cross links. An analysis of SEM pictures has been carried out and results are given in Table 2. As expected after the comparison on the compression behavior, the shortest distance d_{av} is obtained for glass fibers. This is the reason why the initial stiffness of the material is better when compared to the deflection of the beam which depends on the cubic of the distance between two cross links.

This point was not expected as the relative density of the entanglement is higher for the carbon than for the glass. Different surface properties, different reactivity with the epoxy could explain that point.

There are many discussions about the determination of the average distance between fiber contacts. Authors have proposed models to quantify the number of contacts per fiber, in layered structures [21, 22] and in three-dimensional network fibers [23, 24, 25]. The average number of contacts per fiber $<c>$, in the case of 3D random network, is given by equation (1) as a function of L, the fiber length, of f, the volume fraction of fibers (f) and of D, the fiber diameter. Then, the distance, d_{av}, between joints can be obtained (see table 1) by $d_{av} = L/<c>$:

$$\langle c \rangle = 2\frac{L}{D}f \tag{1}$$

If the initial density of the fibers entanglement is the same for all fibers (150 kg/m^3), the volume fraction of the three tested material is not the same. We can notice that the distance between joints obtained by equation (1) is closed from SEM observations for material made of glass and material made of stainless steel fibers. In the case of material made of carbon fibers, there is a large difference between the two values. Further investigations on the microscopic organization of the material elaborated in that study are still necessary. X-ray topography measurement would provide more valuable information [16].

T.W. Clyne et al. developed a simple analytical model based on the bending of inclined individual fiber segments [18, 26, 27], the Young's modulus of the network is given as a function E_f, the Young's modulus of the fibers and of the geometrical parameters of the network (f, D and d_{av}) :

$$E_a = \frac{9 E_f f}{32 \left(\dfrac{d_{av}}{D} \right)^2} \qquad (2)$$

The values obtained with this approach can be compared with the one given by Gibson and Ashby for open cell foams [20]. This second model is also based on beam deflections, but with different boundaries conditions (more constrained). The Young's modulus of the network is then given by:

$$E_a = \frac{3\pi E_f}{4 \left(\dfrac{d_{av}}{D} \right)^4} \qquad (3)$$

Table 2: Comparison of Young modulus obtained during compression tests with the values calculated using equations (2) and (3).

Material	E: Young modulus [MPa]		
	Calculated using Eq (2)	Calculated using Eq (3)	E_{exp}
Network of stainless steel fibers	1.7	1.2	0.6
Network of carbon fibers	7.3	0.8	1.1
Network of glass fibers	7.8	7	8.4

Young modulus has been calculated applying equations (2) and (3) and is given in Table 2. They are compared with the experimental data (initial stiffness of the material). We can notice that the Young modulus obtained using equation (3) is closed to the experimental values for carbon and glass fibers. This point could mean that the epoxy drops at the cross links limit the deflection of fibers. In the case of stainless steel fibers, the difference between experiment and equation (3) could be explained by the plastification of fibers and the poor reticulation of the network.

Tomography data before the compression test and at the beginning of the plateau would be very useful to determine the fibers orientation, the isotropy of the initial entanglement, the average number of contacts per fibers and could help to get a better understanding of the fiber slippage and fiber orientation changes induced by compression.

These observations are now possible and have just been achieved in another study [16]. We plan to get comparable data on our specimens.

CONCLUSIONS

Original materials have been elaborated using fibers networks bonded with epoxy resin. Three different families of fibers have been tested, glass fibers, carbon fibers and stainless steel fibers. The material obtained remains light (200 kg/m^3) as the process developed in this study optimize the quantity of epoxy used. The best stiffness is obtained for glass fibers mainly because the shortest distance between cross-links compared to the carbon case. The initial stiffness of the cross-links architecture seems to follow the model proposed by Ashby.

REFERENCES

[1] P. Ducheyne, E. Aernoudt, P. Meester, J. Mater. Sci. 13(12):2650(1978)
[2] T.W. Clyne, J.F. Mason, Metal. Trans. A 18(8):1519(1987)
[3] F. Delannay, T.W. Clyne, Proceedings of MetFoam'99, 14-16 June, Bremen, Germany (1999)
[4] Y. Yamada, C.E. Wen, Y. Chino, K. Shimojima, H. Hosokawa, M. Mabuchi, Mat. Sci. Forum 419:1013(2003)
[5] A.E. Markaki, V. Gergely, A. Cockburn, T.W. Clyne, Compos. Sci. Technol. 63(16):2345 (2003)
[6] A. Woesz, J. Stampfl, P. Fratzl, Adv. Eng. Mater. 6(3):134 (2004)
[7] M. Delince, F. Delannay, Acta Mater. 52(4):1013 (2004)
[8] L.O., Golosnoy A. Cockburn, T.W. Clyne, Adv. Eng. Mater. 10(3):210(2008)
[9] B.M. Zhang, S.Y. Zhao, X.D. He J. of Quant Spectro & Radiat Transf 109(7):1309(2008)
[10] R. Gustavsson, AB Volvo Patent WO 98/01295, (15th January 1998)
[11] A.E. Markaki, T.W. Clyne, US patent 10/000117, Cambridge University (2001)
[12] A.E. Markaki, T.W. Clyne, Acta Mater. 51(5):1341 (2003)
[13] A.E. Markaki, T.W. Clyne , Acta Mater. 51(5):1351 (2003)
[14] J. Dean et al. Proceedings of ICCS8, Porto, Portugal, edited by Ferreira,: 199.
[15] D. Zhou, W.J. Stronge, Int. J. of Mech. Sci. 47(4-5):775 (2005)
[16] J.P. Masse Ph.D. Institut National Polytechnique de Grenoble, France (2009)
[17] L. Mezeix, C. Bouvet, B. Castanié, D. Poquillon (2008) Proceedings of ICCS8, Porto, Portugal, edited by Ferreira, Portugal: 798 (2008)
[18] T.W. Clyne, A.E. Markaki, J.C. Tan, Compos. Sci. and Technol. 65(15-16):2492 (2005)
[19] L. Mezeix, Material Science Master's degree, Univ. de Toulouse(2007)
[20] L.J. Gibson, M.F. Ashby, Cellular solids: structure and properties. Cambridge University Press(1997)
[21] W.J. Batchelor, J. He, W.W. Sampson, J. of Mater. Sci. 41(24):8337-8381 (2006)
[22] J. He, W.J. Batchelor, R.E. Johnston, J. of Mater. Sci. 42(2):522-528 (2007)
[23] S. Toll Polym. Eng. Sci .38:1337 (1998)
[24] C.T.J.Dodson, Tappi J 79(9):211-216(1996)
[25] A.P. Phillipse, Langmuir 12(5):1127-1133(1996)
[26] A.E. Markaki, T.W. Clyne, Biomaterials 25(19):4805 (2004)
[27] A.E. Markaki, T.W. Clyne, Acta Mater. 53(3):877 (2005)

Mater. Res. Soc. Symp. Proc. Vol. 1188 © 2009 Materials Research Society   1188-LL02-05

Mechanical Properties of Sintered and Non Sintered Stainless Steel Wools: Experimental Investigation and Modeling

J. P. Masse[1,2], O. Bouaziz[2], Y. Brechet[1], L. Salvo[1]
[1]SIMaP, Grenoble INP, SIMAP groupe GPM2 101 rue de la physique, 38402 Saint Martin D'heres.
[2]ArcelorMittal Research, Voie Romaine-BP30320, 57283 Maizières-lès-Metz Cedex, France

ABSTRACT
Entangled fibrous materials have a common point with cellular solids: the architecture is at the millimetric scale. However, they present one extra degree of freedom which is the connectivity on the constituents: while compressing a fibrous structure, the number of contacts between fibers is variable, by contrast to cellular solids where the building lock is the cell. In this respect, fibrous entangled solids can span a whole range of mechanical behavior depending on the possibility offered to fibers to create new contacts: from the felt where they are totally free, to the fully sintered wool where their relative motion is constrained by irreversible contacts. The purpose of this paper is to investigate the mechanical properties of this class of materials, based on micro tomography observations for the number of contacts, and on a physically based model for fiber bending and collective reorganization. The material studied is a stainless steel wool. The properties investigated are the loading curves for sintered and non sintered wools. Qualitative differences introduced by sintering motivate a modification of the classical Toll model, which will be presented together with the experimental results.

INTRODUCTION
Entangled materials, consisting of fibers, exist from natural material (mutton wool, cotton) as well as artificial one (steel wool, glass wool, felts ...). They find applications in thermal insulation, mechanical reinforcement and filtration. However, their mechanical properties are very low compared with traditional cellular materials made from the same constitutive material, and with a similar density. One way to gain a better mechanical behavior is to create permanent bonds between fibers by means of sintering. Conventional sintering consists in the heating the material in a furnace; final density and mechanical properties depend on the sintering time and temperature [1]. The properties of these materials depend on the relative density, their constitutive material, the quality of the contact created, but also on the architectural parameters (fibers orientation, number of contacts). Mechanical properties of these materials (sintered mats of fibers [2] or 3D random bonded fibers [3]) were characterized in compressive test, as it can be done with usual cellular materials [4]. Analytical models [5,6] obtained by dimensional analysis allow to get scaling laws where the compression stress and density are related by a power law relationship. These scaling laws were recently confirmed by discrete 3D simulations [7] and they are well adapted to entangled materials.
The aim of this work is to relate mechanical properties in compression of non sintered and sintered steel wool with their structure (fibers orientation and numbers of contacts), using classical model of the literature.

MATERIALS
Steel wool studied here is provided in form of strip: 5 mm in thickness and 27 cm wide. It is made of ferritic stainless steel AISI 434 fibers. The cohesion of the material as received is only

due to the fibers tangle. The fibers have an aspect ratio of about 400, with a mean equivalent diameter (d_f) of 80 µm and mean length (l_f) of 5.6 cm. Since the average relative density is 2% approximately, we can extract samples 30 mm by 30 mm whose initial relative density (ρ_i) ranging from 1.5% and 3.6%. Conventional sintering allows obtaining samples of final relative density (ρ_f) ranging from 4% to 12%. The densification, defined as the ratio between the initial relative density and the relative density after sintering, is ranging from 2 to 3 for a sintering temperature of 1300°C and from 3 to 4 for T=1400°C. It does not depend on the initial value ρ_i. There is clearly an important effect of temperature on the sintering densification of our material while the holding time has a much smaller influence. From this observation, one can infer that the sintering time must have an impact on the quality of the bonding of the contacts, but probably not on their number. SEM observations confirm the good bonding of the fibers at contact. We will present mechanical properties of the both types of materials, according to the classical Van Wyk – Toll formalism which will be presented in the nest section.

THEORY

The classical model to describe the compression of entangled was derived by van Wyk [6] for 3D random structure and later refined by Toll [5] for 2D random structure. In this model it is assumed that the fibers are straight and disposed homogeneously into the material, and that they are loaded in bending during compression. From particular assumption we can write the stress (σ) for the uniaxial compression as the product of the number of deformation unit per their height per the force applied on them.

$$\sigma = \eta \, hp \tag{1}$$

With η the numbers of deformation unit per unit volume, h the height of the deformation unit and p the force applied on the deformation unit. The number of deformation unit is the number of contact per unit volume (N_{cv}), and the force p is the force is the bending force considering that the span of the beam is equal to the distance between contacts (λ). When take into account this we can express the stress as follow, with ρ the relative density during compression:

$$\sigma \propto \int_0^\rho \frac{N_{cv} \, h^2}{\lambda^3} d\rho \tag{2}$$

The height of the deformation unity is different for 2D and 3D random material. In first case it is proportional with the distance between contact and in the second it is proportional with the diameter of the fibers. The number of contacts in a fibers assembly is well described by the tube model considering rigid straight fibers and homogeneously disposed in the material. For fibers with high aspect ratio, the expression of the numbers of contacts per unit volume (N_{cv}) is:

$$N_{cv} = \frac{16f}{\pi^2 d_f^3} \rho^2 \tag{3}$$

With d_f the diameter if the fibers ad f, a orientation function define by the following equation:

$$f = \frac{1}{N_f^2} \sum_\alpha \sum_\beta \left\| q_\alpha \times q_\beta \right\| \tag{4}$$

Where q is the director vector of the fibers. f= 0.78 for a random 3D fibers assembly, f= 0.63 for 2D and f=0 for aligned fibers. Finally it is possible to integrate equation 2 to obtain:

$$\sigma = kE_f (\rho^n - \rho_0^n) \tag{5}$$

With ρ_0 the packing density, below which no stress is required to compact the material. In the case of a 3D random structure n = 3 and in the case of the 2D random structure n=5.

EXPERIMENT
Mechanical properties
Compression tests were performed on tension/compression machine with a cell force of 2kN, and a LVDT captor. Simple loading were performed on non sintered and sintered steel wool. We obtain from these tests stress-relative density curves. Typical curves obtained for non sintered and sintered steel are represented in figure 1.

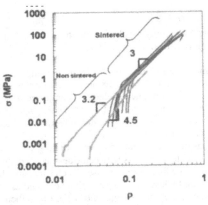

Figure 1. Nominal stress (σ) – relative density (ρ) curves for non sintered and sintered steel with the exponent of the power law

For the two types of materials, the curves are divided into 2 stages. In first step the stress increases rapidly to reach a stress above which, it follows a power law relationship of ρ, as described by Toll's model. For non sintered materials the curves are different depending of ρ_i. The exponent varies from 3.2 for ρ_i=1.5 to 4.5 for ρ_i=3%. For sintered material the stress is larger than for the non sintered steel wool: this observation is to be related to the densification of the samples. For a given density, we notice that there is no influence of the sintering temperature and time on the compressive stress. The exponent of the power law relationship is equal to 3 for all the samples. Toll's model suggests that the stress is related with ρ using structural parameters describing the wool architecture. We will now present a quantitative structural characterization for both sintered and non sintered wools in order to explain the value of the exponent.

Structural characterization
The aim of the structural characterization is to extract the relevant structural parameters of the studied materials. To obtain these parameters X ray microtomography experiments were performed on the both materials in the ESRF (ID 19,Grenoble) and in MATEIS laboratory (INSA Lyon). Two kinds of experiments were led. Simple scan (without load) and in situ compression test (with load). In the first case we can obtain the 3D observations of the initial

materials. In a second case we use a tension/compression machine integrated in the X ray tomography device to obtain 3D observations of the material for different step of deformation. For each observation the pixel size was ranging from 13 to 16 μm. It is sufficient to observe fibers. Figure 2 presents an example of 3D pictures rendering.

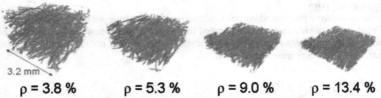

$$\rho = 3.8\,\% \qquad \rho = 5.3\,\% \qquad \rho = 9.0\,\% \qquad \rho = 13.4\,\%$$

Figure 2. Example of a 3D pictures rendering obtained with the same non sintered steel wool at different relative density during in situ compression test.

To extract the structural parameters we identify the skeleton by using Avizo 5 software [12].The skeleton is the wireframe representation of the fibers network, with this transformation the fibers become wire of 1 pixel thickness. The obtained skeleton is composed of segment whose extremities are the cross on the fibers network. We know the coordinates of each segment of the skeleton. It is therefore possible to know their length and, direction in space.
One peculiarity of this method is that we can easily obtain the contacts between fibers. Fig 3 presents different configurations of contacts and the corresponding skeleton identification. Thanks to these observations we detect contacts between fibers as segments in the skeleton shorter than twice the fiver radius, i.e.80 μm in our case. To eliminate double or triple contact (see figure 3 c)), 2 contacts distant by less than 80 μm will be counted as one. Once the contacts are found, we can focus on the other segments representing the fibers themselves. From these segments we can calculate orientation function.

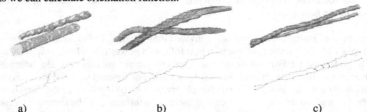

a)   b)   c)

Figure 3. Example of contacts between fibers on the 3D pictures from tomography observation and their associated skeleton

Fig. 4 represents the numbers of contacts and the orientation function f, for the both types of materials for simple scans (without load) and in situ compression tests (with load).
For initial non sintered steel wool sample we observe that the value of f is between the value for a 2D random material and aligned fibers. The fibers are more aligned when ρ_i is high and f reaches a minimal value close to 0.4. The results for in situ compression test show that f reaches the same value during compression, irrespectively of the initial density. The question is how this minimum value is reached, depending on the initial density. Assuming that sintering is mainly a densification, experiments on sintered steel wool without load bring some light to this question.

A sample of non sintered steel wool for which $\rho_i=1.5\%$ (A_{ns}) sintered at 1300°C during 2h has a relative density after sintering of 5% and the value of f is then close to 0.48 (A_s). Another sample for which $\rho_i=2\%$ (B_{ns}) sintered at 1300°C during 2h has a relative density after sintering of 7%, and f is closed to 0.46. We can deduce from these example that the evolution of the orientation during compression is likely to depend on ρ_i.

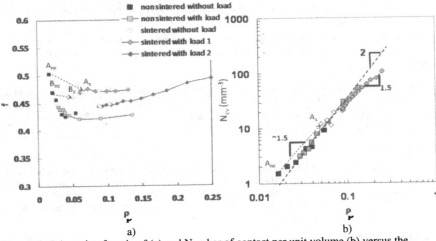

a)   b)

Figure 4. Orientation function f (a) and Number of contact per unit volume (b) versus the relative density for non sintered and sintered with and without load.

According to equation 3 the number of contact depends on the current density, but also on the orientation. So the evolution of the number of contacts during compression depends on ρ_i. Assuming that the number of contacts of the relative density during compression is:

$$N_{cv} \propto \rho^\beta \qquad (6)$$

With this expression we integrate both dependence on orientation and relative density. The exponent is ranging from 1.5 ($\rho_i<3\%\%$)to 2 ($\rho_i=3\% - 10\%$) for non sintered steel wool. For sintered steel wool is value is approximately 1.5 for $\rho>10\%$.

For sintered steel wool, the evolution of the structure is quite different during compression. We observe that the orientation function does not reach an asymptotic value. It seems, there are two steps, until a value of ρ close to 0.12 f is constant and after this value f increase with ρ. In term of numbers of contact this observation leads to a deviation from the standard model giving the number of contacts. It is more difficult to create contacts when initially some of them are bonded. The observed exponent of the power law for sintered wools is close to 1.5.

DISCUSSION

Our goal is to relate mechanical properties with the structure of the steel wool, along the lines of Toll's model. The value of f showed that the considered materials are closed to a 2D structure, so the unit of deformation we must consider is the 2D one of Toll's model: its height h is constant and proportional to the diameter of the fibers. Let's consider now equation 2 and evaluate the parameters of the equation.

The number of contacts has already been evaluated (equation 6) and the distance between contacts is given by: :

$$L = \lambda \propto \rho^{1-\beta}$$ (7)

After the integration of equation 2, we obtain a relation between the exponent β and n:

$$n = 4\beta - 3$$ (8)

The exponent β has been determined experimentally from structural characterization for non sintered and sintered steel wool, and is given in table together with the theoretical and the experimental values of n.

Table I: experimental exponent β and n for the both materials studied and theoretical exponent n.

Materials	Relative density	β	Theoretical n	Experimental n
Non sintered steel wool	1.5%<ρ<3.5%	1.5<β<2	3<n<5	3.2<n<4.5
Sintered steel wool	4%<ρ<12%	β=1.5	n=3	n=3

CONCLUSIONS

The mechanical behavior in compression of non sintered and sintered steel wool has been studied, together with a quantitative 3D characterization of the architecture, with a special emphasis on the number of contacts between fibers. The stress/ density curves are well represented by Toll's power law, and the exponents can be derived accurately from the density dependence of the number of contacts.

ACKNOWLEDGMENTS

Authors want to acknowledge Didier Bouvard and Katrin Beyer for the processing of the sintered samples, Eric Maire and Jerome Adrien for their help on the tomography experiment, and Carine Barbier for her help on the development of the method for contact analysis.

REFERENCES

1. P. Liu, G. He, L.H. Wu : *Materials. Science and. Engineering.* (2008) in press
2. F. Delannay, T.W. Clyne, in: *Metal Foams and Porous Metal Structures* edited by J. Banhart, M.F. Ashby, N.A. Fleck Ver. MIT Publ., 1999
3. A.E. Markaki, V. Gergely, A. Cockburn, T.W. Clyne: Computer. Science. and Technology. Vol. 63 (2003) p 2345
4. L.J. Gibson, M.F. Ashby: *Cellular Solids: Structure and Properties* (Cambridge Univ. Press 1999)
5. S. Toll *Polymer Engineering and Science* Vol. 38, No. 8 (1998) p 1337
6. C.M. van Wyk : *Journal.of the Textile Institute*, vol 37 (1946) p 285
7. D. Rodney, M. Fivel, R. Dendievel: *Physical. Review. Letter.* vol. 95 (2004) p 1.
8. S.S. Panda, V. Singh, A. Upadhyaya, D. Agrawal: *Scripta Materialia.* Vol 54 (2006) p 2179
9 J.P.Masse, L.Salvo, D.Rodney, Y.Bréchet, O.Bouaziz:*Scripta Materialia* vol. 54 (2006) p 1379
10. Raganathan, S. G. Advani : *Journal of Rheology*, 35:1499-1522;1991
11. S. Toll : *Journal of Rheology.*, 37 :123_125, 1993.
12. Mercury Computer Systems, Inc., Avizo, 3D Visualization Framework for Scientific and Industrial Data, 3dviz.mc.com

Processing Challenges

Mater. Res. Soc. Symp. Proc. Vol. 1188 © 2009 Materials Research Society   1188-LL03-01

Metal/Metal Nanocrystalline Cellular Composites

Brandon A. Bouwhuis, Eral Bele, and Glenn D. Hibbard
Department of Materials Science and Engineering, University of Toronto, 184 College Street, Toronto, ON, M5S 3E4, Canada

ABSTRACT

Nanocrystalline electrodeposition can be used to reinforce conventional metallic micro-truss materials and conventional metal foams, creating new types of metal/metal cellular hybrids in which the mechanical performance is controlled by an interconnected network of nanocrystalline tubes. This approach takes advantage of the large strength increase that can be obtained by grain size reduction to the nm-scale and the fact that the electrodeposited material is optimally positioned away from the neutral bending axis of the composite cellular struts or ligaments. This article presents an overview of the potential for structural reinforcement of bending-dominated and stretching-dominated cellular architectures by nanocrystalline electrodeposition.

INTRODUCTION

New regions of material property space can be accessed by combining microstructural design at the nm-scale with architectural design at the μm- or mm-scale. In the first case, large strength increases associated with grain size reduction to below 50 nm have driven extensive research efforts into the development of nanocrystalline materials [e.g. 1-3]. For many potential structural applications, however, the density of a nanocrystalline material is just as important as its strength. In fact, reducing the density is more important than increasing the strength for certain weight specific materials performance indices, and it is especially critical for applying structural nanomaterials in the aerospace and automotive sectors. We have developed a new type of structural nanomaterial wherein the effective density of the parent metal is reduced by more than an order of magnitude by incorporating an internal periodic cellular architecture of open space. In one example a low density cellular nanocrystalline material was created by electroforming nanocrystalline Ni around a rapid-prototyped acrylic photopolymer micro-truss [4]. Micro-truss materials have periodic cellular architectures that are specifically designed to undergo stretching-dominated deformation as opposed to the bending-dominated deformation in conventional open cell metal foams [5-7]. This new cellular nanocrystalline hybrid material combined the structural efficiency of micro-truss architectures with the ultra-high strength that can be achieved by grain size reduction to the nm-scale. Although it played a critical role as a cathode support during the initial stages of nanocrystalline electrodeposition, the polymer core did not contribute significantly to the inelastic buckling resistance of the composite metal/polymer struts and it may therefore be desirable to remove it post deposition by means of chemical dissolution or thermal decomposition [4].

Nanocrystalline electrodeposition can also be used to reinforce cellular metals, creating new types of cellular composites such as metal/metal micro-truss hybrids [8-10] and metal/metal foam hybrids [11,12]. For example, nanocrystalline Ni-Fe alloy sleeve thicknesses of up to 400 μm were used to reinforce an aluminum alloy micro-truss that had been fabricated using a stretch-bend approach; the nanocrystalline sleeves had the effect of increasing the micro-truss strength by up to a factor of twelve and the specific strength by a factor of three [8]. Alternatively, much thinner sleeves can be electroplated to produce structural coatings. In one example, a nanocrystalline Ni coating was designed to provide both corrosion protection and inelastic buckling resistance [10]. Because the ultra-high strength material was optimally located at the furthest distance from the neutral bending axis, only a thin coating of ~50 μm was needed to double the inelastic buckling resistance of the 1.13 mm × 0.63 mm cross-section plain carbon steel struts [10]. Overall, the mechanical performance increase that can be obtained through electrodeposition depends on a complex interaction between the starting cellular architecture and the nanocrystalline sleeves. The following article provides a brief overview of nanocrystalline electrodeposition before contrasting structural reinforcement in bending-dominated and stretching-dominated metal/metal nanocrystalline cellular composites.

NANOCRYSTALLINE ELECTRODEPOSITION

The first systematic studies of nanocrystalline electrodeposition were conducted in the late 1980s [13,14], and the first industrial application was developed in the early 1990s as an in-situ repair technology for Canadian CANDU nuclear steam generators [15]. A wide number of electrodeposited nanocrystalline systems have since been developed, including Fe-, Ni-, Co-, and Cu-based alloys and metal matrix composites (e.g. see review in [16]). One of the most important aspects of nanocrystalline electrodeposition is that it is a non-line-of-sight deposition technique. Coatings or free-standing materials can be produced in virtually any geometry with a wide range of nanostructures. This is the critical characteristic that allows nanocrystalline electrodeposition to be used as a structural reinforcement of cellular metals.

While the effect of grain size reduction on the yield strength of nanocrystalline electrodeposits has been well established, there has been far less consistency in the reported effects of grain size reduction on ductility. The first tensile tests of nanocrystalline electrodeposits were conducted on free-standing Ni that had been mechanically stripped from a supporting titanium cathode [17]. While a large increase in yield strength was obtained (from 86 MPa for a grain size of 10 μm [18] to 910 MPa for a grain size of 10 nm [17]), the tensile elongation to failure was on the order of only ~1% [17]. Similarly discouraging tensile ductilities were reported in a number of subsequent studies [e.g. 19-21]. In addition, two of these studies had clearly demonstrated the presence of sample size effects, where greater failure strengths were seen in smaller coupons [20,21]. Much of this disappointing mechanical behaviour was likely the result of processing flaws, such as hydrogen pit holes and co-deposited bath impurity particulate, introduced into the materials during electrodeposition in small-scale plating systems [22]. However, with improvements to the electrodeposition process and scaling up in pilot plant and full production environments, much higher ductility values are now routinely obtained. For example, a recently published study illustrated that the electrosynthesis method can be used to produce thick nanocrystalline Ni alloy electroforms (e.g. > 2 mm thickness) having grain sizes below 15 nm and exhibiting tensile elongations to failure in excess of 10% without any adverse size effects [23].

It should also be noted that thermal stability is an important issue for nanocrystalline materials. The large excess free energy associated with the intercrystalline region provides a significant driving force for grain growth; for the idealized case of grain boundary curvature-driven growth, the rate of growth is inversely proportional to the grain size. In general, however, growth is retarded by the presence of one or more types of dragging forces which provide a kinetic and/or thermodynamic barrier to grain boundary migration. This allows microstructural design strategies to be implemented that can deliver significantly increased thermal stability. For example, the nanostructure of 10 nm starting grain size nanocrystalline Ni is lost in less than 30 minutes at 543 K [24]. By increasing the starting grain size of Ni to the range of 50 to 100 nm and by micro-alloying with ~3000 ppm of phosphorous, long term thermal stability (on the order of years) can be obtained over the temperature range of 553 K to 593 K [15].

NANOCRYSTALLINE REINFORCED CELLULAR COMPOSITES

Metal/metal composite cellular nanocrystalline materials studied to date include nanocrystalline Ni and Ni-Fe alloy reinforcement of stretch-bend fabricated periodic cellular aluminum micro-trusses (n-Ni/Al [9] and n-NiFe/Al [8]), nanocrystalline Ni reinforcement of stretch-bend fabricated plain carbon steel micro-trusses (n-Ni/steel [10]), and nanocrystalline Ni and Ni-W alloy reinforcement of open cell aluminum foams (n-Ni/Al [12] and n-NiW/Al [11]). In each case, adding the nanocrystalline sleeve significantly increased the initial peak compressive strength. However, the failure strength in these materials is controlled by a complex set of mechanisms. Perhaps the most significant of these involves the relative sequence of architectural collapse (as indicated by the initial peak load in uniaxial compression) and the onset of fracture in the electrodeposited sleeves. For the case of n-NiFe/Al [8] and n-Ni/Al [9] micro-trusses, sleeve fracture occurred before the peak strength and could be seen as a series of sharp valleys in a plot of the tangent modulus as a function of strain [8,9]. Similarly, the n-NiW/Al foam hybrids exhibited a number of small load drops before the peak compressive strength [11], which suggests that early sleeve fracture occurred in this system too. On the other hand, there was a smooth transition through the peak stress for both the n-Ni/steel micro-truss [10] and n-Ni/Al foam [12] hybrids and no evidence of crack formation was seen until after the peak stress had been reached. Whether sleeve fracture happens before the peak load or not depends on variables such as the ductility of the electrodeposited sleeves, the adhesion between the sleeve and the starting cellular core, and the architecture of the cellular pre-form. In terms of modeling the peak compressive strength, the simplest scenario occurs when both the nanocrystalline sleeve and the starting cellular core undergo the same overall failure mechanism, i.e. for the case when nanocrystalline sleeve fracture occurs after the peak compressive stress has been reached. The following sections contrast the examples of n-Ni/steel micro-truss [10] and n-Ni/Al foam [12] hybrids, illustrating the relative significance of nanocrystalline structural reinforcement on stretching-dominated and bending-dominated cellular architectures.

The starting pre-form for the n-Ni/Al foam hybrids was open cell AA6101 aluminum foam having a relative density of $\rho \approx 7.2\%$, nominal pore size of 1.27 mm (20 pores per inch) that had been purchased from ERG Aerospace (of Oakland, CA) [12]. The starting pre-form for the n-Ni/steel micro-truss hybrids was perforated AISI-SAE 1006 low carbon steel sheet that had been stretch-bend fabricated into a pyramidal micro-truss core (stretch-bend fabrication details in [25]); the as-fabricated steel core had a relative density of 5.9% and internal strut angle of $\omega =$

35° [10]. In both cases, nanocrystalline nickel reinforcement was conducted by electrodeposition using a modified Watt's bath containing nickel sulfate, nickel chloride, boric acid, and saccharin with a consumable Ni anode at a pH of 2.5, after [26-28]. The thickness of the nanocrystalline sleeves was controlled by the electrodeposition time. The nanocrystalline Ni grain size was determined from reference coupons by X-ray diffraction (XRD) characterization using Co-K$_\alpha$ radiation (λ = 1.79 nm), Figure 1. An average grain size of 16 nm was measured from the diffraction peak broadening (using the Scherrer relationship [e.g. 29]), which is typical of n-Ni produced by pulse current electrodeposition, e.g. [28,30]. The n-Ni thickness distribution on the electrodeposited foams was characterized by scanning electron microscopy (SEM) of polished cross-sections. Vickers micro-hardness measurements (0.49 N load, 10 s dwell time) were conducted on the nanocrystalline sleeves of the n-Ni/Al foam (472 ± 26 HV) and n-Ni/steel micro-truss hybrids (475 ± 14 HV). Each type of hybrid was also tested in uniaxial compression. The n-Ni/steel micro-truss hybrids were tested using confinement plates (i.e. recessed channels in steel compression plates that rigidly lock the truss-core nodes in place, details in [31]). This test method can be used to simulate the mechanical performance that stand-alone truss cores would exhibit in a sandwich structure [31].

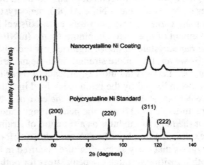

Figure 1. X-ray diffraction patterns of electrodeposited nanocrystalline Ni (λ = 0.179 nm) and polycrystalline Ni powder standard.

A central issue for the n-Ni/Al hybrid foams was the coating thickness uniformity, since this directly determined the spatial distribution of structural reinforcement. A nominal thickness (t_{n-Ni}^{nom}) for the nanocrystalline Ni sleeves can be calculated by assuming a uniform distribution of electrodeposited mass (m_{n-Ni}) over the specific surface area (S) of the starting cellular architecture according to the relationship:

$$t_{n-Ni}^{nom} = \frac{m_{n-Ni}}{SV\rho_{Ni}} \qquad (1)$$

where ρ_{Ni} is the density of Ni, and V is the volume of the cellular pre-form. In the case of the n-Ni/Al foams, nominal n-Ni thicknesses ranged from 26 μm to 72 μm [12]. However, electromagnetic shielding issues induced by the outer ligaments [e.g. 32] resulted in a thickness gradient through the height of the starting foam, which can be seen in Figure 2. For all n-Ni/Al foam samples, the thickness of the n-Ni sleeves was smallest and most uniformly distributed within the middle of the electrodeposited foam.

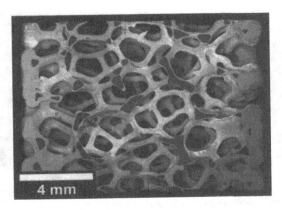

Figure 2. SEM micrograph of a n-Ni/Al foam hybrid showing the distribution of nanocrystalline Ni reinforcement (light) over the aluminum core (dark).

The cross-section of a typical n-Ni reinforced aluminum ligament from the middle of the foam can be seen in Figure 3a; the n-Ni thickness on the mid-foam height ligaments increased linearly with overall hybrid density and was approximately 40% of that of the nominal thickness value [12]. It should be noted that some degree of thickness non-uniformity in the nanocrystalline foam reinforcement may be beneficial if the structural objective of the hybrid foam is to resist externally applied bending loads; in this situation, having the majority of the nanocrystalline reinforcement located near the outer surfaces may provide enhanced weight specific bending resistance compared to uniformly coated foam cores. On the other hand, the n-Ni reinforcement on the plain carbon steel micro-truss was more uniform, largely the result of the simpler pyramidal cellular architecture. Nominal n-Ni coating thicknesses on the plain carbon steel micro-truss ranged from 20 μm to 60 μm (typical strut cross-section shown in Figure 3b). The mid-strut thicknesses in these hybrids were within 10% of the nominal coating thickness. Note, however, that the electrodeposited sleeve thickness alone is not a meaningful point of comparison between the hybrid foam and micro-truss studies. At 0.32 mm^2/mm^3, the specific surface area of the starting plain carbon steel micro-truss was considerably smaller than that of the starting aluminum foam (1.19 mm^2/mm^3); this meant that even though the nominal n-Ni coating thicknesses were comparable, the density increase of the n-Ni/Al foam hybrids was much greater.

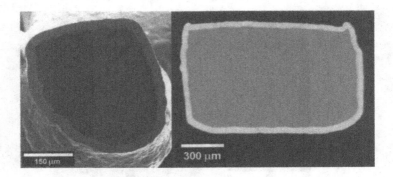

Figure 3. SEM micrographs showing typical composite cross-sections from a n-Ni/Al foam ligament (left) and a n-Ni/steel micro-truss strut (right).

Figure 4 presents typical uniaxial compression stress-strain curves for both the n-Ni/Al foams and the n-Ni/steel micro-trusses. Significant increases to the peak strength and compressive modulus were seen with increasing electrodeposit sleeve thickness in each system. The thinnest nanocrystalline sleeves (corresponding to $t_{n-Ni}^{mid} = 13.0$ μm) more than doubled the peak strength of the hybrid foams, while the thickest n-Ni sleeves (corresponding to $t_{n-Ni}^{mid} = 28.8$ μm), increased the peak strength by a factor of 5.2 (from 1.26 to 6.54 MPa). Note that compression tests were performed along the long axis of the cross-section in Figure 2 and that architectural failure in all cases was initially confined to the mid-height region of the foam where there was the least n-Ni structural reinforcement [12]. For the n-Ni/steel micro-trusses, the electroplated samples underwent the same inelastic buckling failure mechanism as the uncoated samples and a ~60 μm thick coating resulted in a 120% increase in peak strength. Conventional open-cell aluminum foams typically fail at an approximately constant plateau stress once the fully plastic bending moment of the ligaments has been reached [33]. In contrast, the drop in strength seen for the n-Ni/Al hybrid foams after the peak corresponded to the progression of sleeve fracture through the composite ligaments [12]. In the case of the n-Ni/steel hybrids, sleeve fracture did not occur until the local minimum of the stress-strain curve was reached, i.e. corresponding to the point when the inelastically buckled composite struts made contact with the compression platens [10].

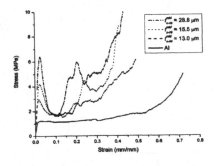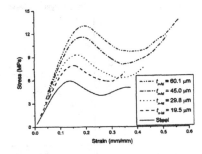

Figure 4. Typical uniaxial compression stress-strain curves of the n-Ni/Al hybrid foams (left) and n-Ni/steel hybrid micro-trusses (right).

A simple approach to modeling the peak strength of the n-Ni/Al foam and n-Ni/steel micro-truss hybrids is to consider the strength increase provided by the nanocrystalline Ni reinforcement ($\Delta\sigma$) in terms of the predicted behaviour of a hollow tube nanocrystalline Ni foam [12] or micro-truss [10]. This approach effectively treats the electrodeposited samples as a two-component composite cellular material. However, unlike the case for fibre or particle reinforced composites, where the addition of the reinforcing phase necessarily displaces an equivalent volume fraction from the matrix, the components of the present composite are interpenetrating structures in which the effective area fraction of each component is essentially constant regardless of the n-Ni sleeve thickness. An initial estimate for the mechanical performance can therefore be taken by using an iso-strain approach in which the aluminum foam core and plain carbon steel micro-truss suffer the same strain during loading as the conformal n-Ni sleeves.

The predicted failure strength for a hollow tube nanocrystalline Ni foam can be estimated from an idealized model based on the fully plastic bending moment of a hollow tube foam [6,33], giving the expression [12]:

$$\sigma_{n-Ni}^{Foam} = \frac{C\sigma_{YS,n-Ni}\left((d+t_{n-Ni})^3 - d^3\right)}{6L^3} \quad (2)$$

where the parameter C describes the end condition (8 for a fixed end-fixed end ligament point loaded at the mid-span [33]), $\sigma_{YS,n-Ni}$ is the yield strength of the electrodeposited nanocrystalline Ni (900 MPa [34]), d is the ligament diameter of the starting aluminum foam, and t_{n-Ni} is the thickness of the nanocrystalline sleeve. Figure 5 plots the experimentally measured strength increase against the density increase for the n-Ni/Al hybrid foams. Also shown is the predicted strength of the conformal hollow tube nanocrystalline Ni foam. Using the coating thickness from the middle of the foam (i.e. the weakest zone where failure first occurred) results in good agreement between model and experimental measurement. It should be noted that simply using the nominal coating thickness instead of the actual coating thickness from the middle of the foam significantly over-predicts the peak strength by a factor of ~2 to 3 times [12].

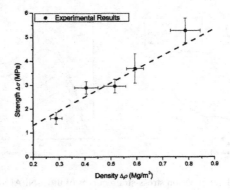

Figure 5. Experimentally measured strength increase ($\Delta\sigma$) as function of the density increase ($\Delta\rho$) for the n-Ni/Al foams. Also shown is the predicted strength increase for a conformal hollow-tube n-Ni foam based on Equation 2.

A similar approach can be taken to determine the predicted strength of a hollow tube nanocrystalline Ni micro-truss; the idealized strength for a pyramidal micro-truss [35] where inelastic buckling is the strength controlling failure mechanism (see [36]), can be expressed as [10]:

$$\sigma_{n-Ni}^{truss} = \frac{k^2 \pi^2 E_t I \rho_R \sin^2 \omega}{AL^2} \tag{3}$$

where E_t is the tangent modulus of the nanocrystalline Ni ($E_t = \partial\sigma/\partial\varepsilon$), A is the cross-sectional area of the supporting member, I is the moment of inertia, and k describes the rotational stiffness of the column. The predicted inelastic buckling strength as a function of density for a hollow tube nanocrystalline Ni micro-truss is plotted in Figure 6, using the limiting cases of pin ($k = 1$) and rigid ($k = 2$) end constraints. Note that the actual end constraints of conventional micro-truss materials are expected to fall within these bounds [e.g. 37]. Figure 6 shows that there is reasonable agreement between the predicted and experimentally measured strength increase for the n-Ni/steel micro-trusses.

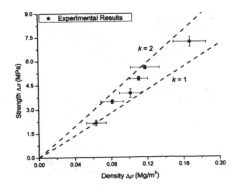

Figure 6. Experimentally measured strength increase ($\Delta\sigma$) as function of the density increase ($\Delta\rho$) for the n-Ni/steel micro-trusses. Also shown is the predicted strength increase for a conformal hollow-tube n-Ni micro-truss based on Equation 3.

To a first approximation, the plastic bending moment of a hollow tube nanocrystalline foam [12] and the inelastic buckling strength of hollow tube nanocrystalline micro-truss [10] can reasonably account for the experimentally measured strength increase provided by nanocrystalline electrodeposition. While issues such as coating uniformity and nanocrystalline sleeve fracture play an important role in determining the failure strength, the most significant effect is whether the network of nanocrystalline tubes undergoes bending-dominated or stretching-dominated deformation. This can be seen in Figure 7, where the strength increase provided by the nanocrystalline sleeves ($\Delta\sigma$) is plotted against the density increase ($\Delta\rho$) for the hybrid nanocrystalline systems studied to date. When electrodeposited on an open-cell foam, the effective specific strength ($\Delta\sigma/\Delta\rho$) of the nanocrystalline Ni network of tubes ranged from ~6 to 7 MPa/(Mg/m^3). Using a nanocrystalline Ni-W alloy that had a higher hardness (5 to 7 GPa [11] vs. 4.7 GPa in the n-Ni/Al foams), resulted in a larger effective specific strength (~8 to 14 MPa/(Mg/m^3)). On the other hand, the specific strength increase for nanocrystalline struts undergoing inelastic buckling failure over the 0.1 to 1 Mg/m^3 density range was between ~29 to 52 MPa/(Mg/m^3) [4,8-10]. Nanocrystalline reinforcement is therefore most effective on stretching-dominated cellular architectures; it is anticipated that even larger property enhancements will be attainable when higher strength nanocrystalline alloys are electrodeposited on more structurally efficient micro-truss cellular architectures.

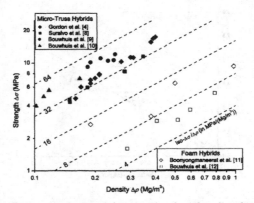

Figure 7. Experimentally-measured strength increase $\Delta\sigma$ plotted against the density increase $\Delta\rho$ for metal/metal hybrid nanocrystalline micro-trusses [4,8-10] and foams [11,12].

CONCLUSIONS

Nanocrystalline electrodeposition can be used to reinforce conventional cellular metals, creating new types of metal/metal cellular composites. For the case when the peak load of the cellular architecture is reached before nanocrystalline sleeve fracture, a composite cellular material approach can be taken in which the strength increase is determined by the predicted behaviour of a conformal hollow tube nanocrystalline cellular material. When electrodeposited on a micro-truss pre-form, the nanocrystalline sleeves failed by inelastic buckling; when electrodeposited on an open-cell foam, they failed by plastic bending. The magnitude of the structural reinforcement attainable for a given mass of electrodeposited material is therefore largely determined by whether the composite cellular material undergoes stretching-dominated or bending-dominated architectural collapse.

REFERENCES

1. H. Gleiter, *Acta Mater.* **48**, 1 (2000).
2. M.A. Meyers, A. Mishra, D.J. Benson, *Prog. Mater. Sci.* **51**, 427 (2006).
3. C.C. Koch, *J. Mater. Sci.* **42**, 1403 (2007).
4. L.M. Gordon, B.A. Bouwhuis, M. Suralvo, J.L. McCrea, G. Palumbo, and G.D. Hibbard, *Acta Mater.* **57**, 932 (2009).
5. M.F. Ashby and Y.J.M. Brechet, *Acta Mater.* **51**, 5801 (2003).
6. M.F. Ashby, *Phil. Trans. R. Soc. A* **364**, 15 (2006).
7. H.N.G. Wadley, *Phil. Trans. R. Soc. A* **364**, 31 (2006).
8. M. Suralvo, B.A. Bouwhuis, J.L. McCrea, G. Palumbo, and G.D. Hibbard, *Scripta Mater.* **58**, 247 (2008).

9. Bouwhuis BA, Ronis T, McCrea JL, Palumbo G, and Hibbard GD. In: Proceedings of CellMet 2008 conference (Dresden, Germany), in press.

10. B.A. Bouwhuis, T. Ronis, J.L. McCrea, G. Palumbo, and G.D. Hibbard, *Comp. Sci. Tech.* **69**, 385 (2009).

11. Y. Boonyongmaneerat, C.A. Schuh, D.C. Dunand, *Scripta Mater.* **59**, 336 (2008).

12. B.A. Bouwhuis, J.L. McCrea, G. Palumbo, and G.D. Hibbard, *Acta Mater.*, in press.

13. G. McMahon and U. Erb, *Microstruct. Sci.* **17**, 447 (1989).

14. G. McMahon and U. Erb, *J. Mater. Sci. Lett.* **8**, 865 (1989).

15. F. Gonzalez, A.M. Brennenstuhl, U. Erb, W. Shmayda, and P.C. Lichtenberger, *Mater. Sci. Forum* **225-227**, 831 (1996).

16. U. Erb, K.T. Aust, and G. Palumbo, in Nanostructured Materials: Processing, Properties and Applications, 2nd Ed. C.C. Koch editor (William Andrew Inc., Norwich, 2007).

17. N. Wang, Z. Wang, K.T. Aust, and U. Erb, *Mater. Sci. Eng.* **A237**, 150 (1997).

18. A.W. Thompson, *Acta Metall.* **25**, 83 (1977).

19. A.F. Zimmerman, G. Palumbo, K.T. Aust, and U. Erb, *Mater. Sci. Eng.* **A328**, 137 (2002).

20. F. Dalla Torre, H. Van Swygenhoven, and M. Victoria, *Acta Mater.* **50**, 3957 (2002).

21. G.J. Fan, L.F. Fu, D.C. Qiao, H. Choo, P.K. Liaw, and N.D. Browning, *Scripta Mater.* **54**, 2137 (2006).

22. I. Brooks, P. Lin, G. Palumbo, G.D. Hibbard, and U. Erb, *Mater. Sci. Eng.* **491**, 412 (2008).

23. H.S. Wei, G.D. Hibbard, G. Palumbo, and U. Erb, *Scripta Mater.* **57**, 996 (2007).

24. U. Klement, U. Erb, A.M. El-Sherik, and K.T. Aust, *Mater. Sci. Eng. A* **203**, 177 (1995).

25. B.A. Bouwhuis and G.D. Hibbard, *Metall. Mater. Trans. A* **39**, 3027 (2008).

26. U. Erb and A.M. El-Sherik, US Patent No. 5 353 266 (October 1994)

27. U. Erb, A.M. El-Sherik, C.K.S. Cheung, and M.J. Aus, US Patent 5 433 797 (July 1995)

28. A.M. El-Sherik, and U. Erb, *J. Mater. Sci.* **30**, 5743 (1995).

29. B.D. Cullity, *Elements of X-ray Diffraction*, 2nd ed. (Addison-Wesley, Don Mills, 1978).

30. A.M. El-Sherik, U. Erb, G. Palumbo, and K.T. Aust, *Scripta Mater.* **27**, 1185 (1992).

31. B.A. Bouwhuis, E. Bele, and G.D. Hibbard, *J. Mater. Sci.* **43**, 3267 (2008).

32. M. Paunovic, M. Schlesinger, and R. Weil, in *Modern Electroplating*, 4th ed., edited by M. Schlesinger and M. Paunovic, (Wiley Interscience, New York, 2000).

33. M.F. Ashby, A.G. Evans, N.A. Fleck, L.J. Gibson, J.W. Hutchinson, H.N.G. Wadley, *Metal Foams – A Design Guide*, (Butterworth-Heinemann, Oxford, 2000).

34. Integran Technologies Inc. (private communication).

35. V.S. Deshpande and N.A. Fleck, *Int. J. Solids Struct.* **38**, 6275 (2001).

36. F.R. Shanley, *Strength of Materials*, (McGraw-Hill, New York, 1957).

37. D.J. Sypeck and H.N.G. Wadley, *Adv. Eng. Mater.* **4**, 759 (2002).

Mater. Res. Soc. Symp. Proc. Vol. 1188 © 2009 Materials Research Society   1188-LL03-06

Parametric Studies in the Processing of a Thermal Wicking Material Using Image Analysis

Stephanie J. Lin[1] and Jason H. Nadler[1]

School of Materials Science and Engineering

Georgia Institute of Technology

Atlanta, GA, 30332 U.S.A

ABSTRACT

A heat pipe is a device that transports heat against gravity using a wicking material and evaporation-condensation cycle. In these systems a thermal wick moves fluid from the cool region of a heat pipe to the hot region, where evaporative cooling occurs. Due to the operating demands of a thermal wick, several microstructural features are integral to the performance of the wick: capillary radii, specific surface area and permeability. Measuring these properties of a thermal wick (capillary radii, specific surface area and permeability) is difficult, therefore image analysis methods of quantification of the critical properties of a thermal wick has been developed. However, the microstructure of a thermal wick contains semicontinuous pores, therefore connectivity of pores cannot be assumed during quantification of the critical properties.. Two processing parameters, sacrificial template particle size and sintering temperature, were varied during the thermal wick synthesis. Quantification of the critical properties of the thermal wick was performed using the newly developed method. The newly developed method was able to detect the an increase in the pore connectivity as the sintering temperature decreased, and an increase in the connectivity as the sacrificial template particle size decreased. The newly developed method was also able to describe the size distribution of individual pores as well as the hydraulic resistance and orientation of individual pores as well as estimate the porosity and true specific surface area of the different samples.

INTRODUCTION

A heat pipe is a device that transports heat against gravity using a wicking material and evaporation-condensation cycle.[1] In these systems, heat is removed through a closed loop of liquid wicking, evaporation and condensation. A thermal wick, constructed of layers of fine gauze[1], for example, must exhibit high intrinsic thermal conductivity, efficient capillary flow and evaporative transport over a wide area through a limited thickness. Although these mechanisms can be optimized through characterization and design on several length scales, performance will arguably be dominated by the wick's micro-/mesoporous architecture.

The capillary radii of pores and specific surface area in the wick microstructure affect the permeability of microstructure, thus affecting the critical properties of the thermal wick. A material with an open cell porosity would provide the microstructural features necessary to obtain the critical properties necessary for the operation of a thermal wick. To achieve the complex microstructure, sacrificial template methods, similar to those employed in macroporous ceramic material synthesis are used[2]. The resulting material exhibits a tailored open porosity;

Thus, a method needs to be developed that reliably and efficiently links quantitative changes in the wick's critical structural features to processing parameters.

Typically, the critical properties of a thermal wick (capillary radius, effective surface area, permeability) are properties that are difficult to measure. Therefore methods of quantitative analysis of continuous porous structures such as Berea sandstone[3] and polystyrene beads[4] using image analysis have been developed. However the methods used in quantitative analysis of a continuous porous structure are not valid for semicontinuous porous microstructure of a thermal wick. Therefore a method of quantifying the critical properties of a semiporous material must be developed.

Methods of pore space partitioning and stochastic modeling of 3D microstructure using image analysis rely on the image processing software to aid with the quantitative analysis of the microstructure. The k-means clustering algorithm is a pixel-based segmentation of image in which each pixel is assigned to a cluster based upon the pixel's grayscale value and proximity to a randomly initialized cluster's centriod [6], can be used as unbiased method of converting grayscale images to binary images. Also many algorithms used for pattern recognition and path finding are applied to SEM images to quantify the geometry and topography of a microstructure. The Euclidian distance transformation (see fig 3.a) is the transformation of binary image into the map of the Euclidian distances between edge to edge. Figure 1 illustrates how the EDM algorithm transforms the image. The watershed algorithm is a method of separating particles. The algorithm finds the ultimate eroded points (UEP) of a binary image, and then dilates the UEP until the edge of a particle or the edge of another UEP is reached. Skeletonization, algorithm used in pattern recognition for robots, has been used to detect differences in the geometry and connectivity of two stochastic reconstructions of Berea sandstone with the same porosity and same correlation functions .[3] In a likewise manner, 3-d replicas of pore networks have been created using quantifying low order statistical information from image processing techniques performed on 2-d micrographs of porous media.[4]

In the study, a method was developed to estimate the permeability of a thermal wick material using image analysis. Two processing parameters, sintering temperature and sacrificial template particle size, were changed to create observable differences within the microstructure of the thermal wick samples. The newly developed quantitative image analysis method was tested to see if it could detect differences within the microstructure.

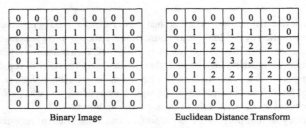

Binary Image　　　　　　Euclidean Distance Transform

Figure 1. Kernel of the distance transformation of a binary image.

76

EXPERIMENTAL DETAILS

Thermal Wick Synthesis

Template particle size was confirmed using laser light scattering. Sets of template particles were mechanically mixed into a liquid slurry, containing 24% wt. ceramic precursor. Samples from each of these batches batch were doctor bladed onto a PTFE substrate at a consistent thickness, air dried and heat treated. To change the surface area and capillary radii of the pores, samples with average sacrificial template particles sizes of 196.1, 94.38 and 18.65 μm, were added at a constant volume fraction of 65%, and heat treated at 700°C. Samples for the sintering study, containing 90% vol. sacrificial template particles of 18.65 μm, were heat treated at 600, 650, and 700°C. Doctor bladed heat treated samples were submerged vertically in commercially available epoxy and vacuum infiltrated. Following curing and polishing, backscattered electron images were obtained using a Hitachi S4100 with an accelerating voltage of 15kV.

Image Processing

Using ImageJ v1.41 software environment, a program was written to automate the binarization of the BSE/SEM cross sectional image using a automated K- means clustering segmentation algorithm with 2 clusters and 48 initial seeds and a center cluster tolerance of 0.00010. Using the binary image, other image processing techniques were used to process the images so that information used for calculating the porosity, specific surface area and hydraulic resistance(as a method of estimation of the permeability of microstructure) of the microstructure could be obtained. Since the porous material consists of two semi continuous phases, many pores located throughout solid phase, (fig 2.a), are not seen in conventional image analysis techniques, such as tangent counting and particle analysis. Therefore conventional image analysis techniques were insufficient for the determination of the specific surface area. In a new method, illustrated in fig. 2, a watershed algorithm is performed on a binary image (see fig 2.b), and then the watershed image is removed from the binary image (fig 2.c)..

Figure 2. Image processing for the calculation of specific surface area (black is the semi continuous phase and white is the pore space). A) binary SEM image B)watersheded binary image C)watershed effect of binary image

The Euclidian distance transformation was applied to the segmented SEM images to produce a Euclidian distance map (EDM) of binary images.. Skeletonization, was performed on the binary image to produce a morphological representation of the original image. The EDM was overlaid onto the skeleton to produce an EDM skeleton (seen in figure 3.b). The skeleton of the binary image was eroded so that only nodes, points with 2 nearest neighbors, remain (see figure 3.c). The nodes are then removed the EDM skeleton, and the data from the image is then quantified and analyzed.

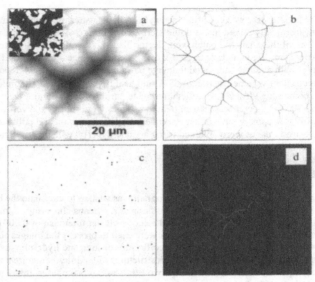

Figure 3.Example of the Image Processing techniques used for permeability measurements. a) EDM transformation of a binary image. b) EDM skeleton of binary image. c) Nodes of the skeleton. D) EDM skeleton with nodes removed.

Quantitative Analysis

To calculate the porosity, the stereological relationship between average area fraction and volume fraction (equation 1) was used to estimate the porosity of the material where $\langle A^{pore} \rangle$ is the average area fraction of the pore space and V^{pore} is the volume fraction of pore space (porosity). Area fractions of pore space (black pixels/ total pixels) corresponding to a sample were used to obtain an average area fraction.

$$\langle A^{pore} \rangle = V^{pore} \tag{1}$$

To calculate the specific surface area, area(in pixels) and perimeter(in pixels) measurements were obtained from the particle analysis of the binary image (fig. 2.a), watersheded binary image (fig. 2.b) and watershed effect (fig. 2.c). Equation 2 was used to calculate the perimeter density. The perimeter density (surface area/ total area) of a planar cross section(S^{planar})can be correlated to the true specific surface area(S^{true}) of the 3-d microstructure.[7]

$$S_{planar} = \frac{\pi}{4} S_{true} \tag{2}$$

Previous studies have correlated the fluid resistance throughout pores to permeability[3-5,8], therefore morphological descriptors of the pore structure are used to quantify the fluid resistance of the structure; the average grayscale value of individual skeleton branches are used to quantify the average pore radii (hydraulic radius) with the microstructure. The length of the individual skeleton branches will be used in the calculation of the fluid resistance of the individual pores, equation (3), where r is the average pore radii of the skeleton branch, L is the length of the skeleton branch and G is the hydraulic resistance.

$$2\pi r \cdot L = G \tag{3}$$

RESULTS & DISCUSSION

Porosity and Surface Area

The results from the porosity and surface area measurements are summarized in table 1. For the sacrificial template size samples, the results indicate that as the sacrificial template particle size decreases, the porosity of the thermal wick increases and the effective surface area of the thermal wick increases. These results are consistent with theory. As the particle size of the sacrificial template particles size approaches the particle size of the matrix precursor, the distribution of the sacrificial template particles within the composite slurry is expected to be more uniform. For the sintering temperature samples, the results indicate that as the heat treatment temperatures decreases, the porosity of the thermal wick increases and the effective surface area of the thermal wick increases. The decrease in the porosity and the decrease in the effective surface area is expected as the heat treatment temperature increase because as the more sintering occurs. SEM images of the samples at the different heat treatment temperatures also indicate that more sintering occurs at the higher heat treatment temperatures.

Table 1. Summary of the Porosity and Specific Surface Area Measurements

	Template Particle Size			Sintering Temperature		
	196.1μm	*94.38μm*	*18.55μm*	*600°C*	*650°C*	*700°C*
V^{pore}	26.57% (±3.56%)	34.75% (±4.97%)	46.65% (±4.63%)	84.19% (±3.26%)	82.23% (±2.46%)	74.08% (±3.34%)
S^{planar}	$0.41\mu m^{-1}$ (±0.07)	$0.53\mu m^{-1}$ (±0.03)	$0.62\mu m^{-1}$ (±0.06)	$0.77\mu m^{-1}$ (±0.04)	$0.74\mu m^{-1}$ (±0.03)	$0.73\mu m^{-1}$ (±0.04)

Pore Structure Descriptions and Hydraulic Resistance Measurements

Table 2 summarizes the results of the pore radius and pore length measurements. Among the samples that varied sacrificial template particle size, the 18.55μm sample exhibited the largest mean pore length (4.82 μm) and the largest skew which indicates that the microstructure of the 18.55 μm sample contains the longest pores, and thus the greatest connectivity. The results also indicate that the connectivity of pore structure of 94.38μm sample is greater connectivity than of pore structure of the 196.1 μm sample although the microstructure of the 196.1μm sized sacrificial template particles contained the biggest pores. Among the samples that varied the sintering temperature , the results indicate that the microstructure of the 700°C sample contains the longest pores, thus the greatest conductivity. The results also suggest that the connectivity of the pore structure of increases as sintering temperature decreases from 650 to 600°C. The results also indicate that the pores within the 600°C sample larger and more uniform than the pores of the 650°C sample.

Table 2. Summary of the distribution of pore radius and pore lengths

	Template Particle Size			Sintering Temperature		
	196.1µm	*94.38µm*	*18.55µm*	*600°C*	*650°C*	*700°C*
Skew of r^{pore}	1.04	1.21	1.10	0.94	1.18	1.60
$< r^{pore} >$	0.59µm	0.53µm	0.43µm	0.49µm	0.46µm	0.50µm
Skew of l^{pore}	1.66	1.70	1.74	1.35	1.35	1.05
$< l^{pore} >$	3.88 µm	3.55 µm	4.82 µm	3.50 µm	3.26 µm	4.13 µm

Figure 4 describes hydraulic resistance at a given orientation for the thermal wick samples. The percentage of hydraulic resistance for a given orientation provides information regarding a fluid's resistance to flow in a particular direction . Therefore the higher percentage of hydraulic resistance for a given angle, the greater the resistance to fluid flow in that direction. From the polar graphs(fig.4), the sintering temperature samples show a high percentage of hydraulic resistance at an orientation of 100° and 180° , suggesting that fluid is more resistant flow thru orientations of 100° and 180° ;the sacrificial template size samples of 196.1µm and 94.38µm show a high percentage of hydraulic resistance at an orientation of 50° and 180° , suggesting that fluid is more resistant flow thru orientations of 50° and 180° the three heat treatment temperatures exhibited the same variance for the orientations. This indicates that the degree of isotropy within the heat treatment samples is the same. The 18.55µm sample exhibits evenly distributed percentage of hydraulic resistance values throughout all orientations, indicating there is no preferential direction for fluid flow in the microstructure of the 18.55 µm sample. For the sacrificial template particle samples, the variance of the hydraulic resistance values based on orientation (0.01%) of the 18.55µm

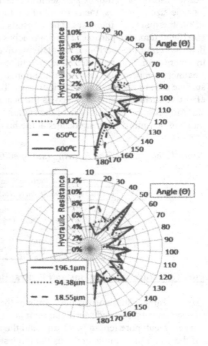

Figure 4. Graph of normalized effective hydralic resistance ploted against the angle(in degrees) for the different processing parameters.

sample was the smallest of the three variances ($196.1\mu m - 0.03\%$, $94.38\mu m$-0.05). These results indicate that the $18.55\mu m$ sample exhibits the greatest degree of isotropy of the three different sacrificial templates, which are consistent with what is expected. The smaller particle size of the sacrificial template particles allow for a more uniform distribution of sacrificial template particles within the green body of the sample, thus creating a more isotropic microstructure prior to heat treatment. However, the small variance within the $18.55\mu m$ sample could be attributed to the thickness of the sample that was used ; the thickness of the $18.55\ \mu m$ was $\sim 100\ \mu m$ thick where as the thickness of the $196.1\ \mu m$ and the $94.38\ \mu m$ sample was 300- $400\ \mu m$ thick.

CONCLUSIONS/FUTURE WORK

A method was developed to quantify the critical properties of thermal wick using image analysis. The method involved the calculation of porosity, specific surface area and statistical analysis of average pore radii and pore lengths from BSE/SEM images. Results from the porosity and surface area measurements corroborated with the expected results. The statistical analysis of pore radii and pore lengths provided information regarding the difference in the microstructures are consistent with the expected results. Further development of the fluid resistance calculation methods is necessary. Permeability experiments will be performed to correlate the differences in fluid resistance with differences in permeability.

REFERENCES

1. D Reay, Peter Kew *Heat Pipes Theory, Design and Applications* 5th ed.(Elsevier, Burlington, 2006)

2. A.R. Studart *J.Am. Ceram. Soc* **89** (6) 1771(2006)
3. Z Liang et al. *Chem Eng. Sci.***55** 5247 (2000)
4. J. Koplik *J Appl Phys.,***56** 3127 (1984)
5. IJ Plugins:Clustering, http://ij-plugins.sourceforge.net/plugins/clustering.
6. J Berryman., *J. Appl. Phys.* **83** (3) 1685 (1998).
7. Lock et al. *J Appl. Phys.*, **92** (10) 6311(2000).
8. Z liang et al. J. Colloid Interface Sci. 221 13(2000).

Mater. Res. Soc. Symp. Proc. Vol. 1188 © 2009 Materials Research Society

Comparison of Two Metal Ion Implantation Techniques for Fabrication of Gold and Titanium Based Compliant Electrodes on Polydimethylsiloxane

Muhamed Niklaus, Samuel Rosset, Philippe Dubois, Herbert R. Shea
Ecole Polytechnique Fédérale de Lausanne (EPFL), Institute of Microengineering (IMT),
Microsystems for Space Technologies Laboratory (LMTS), Rue Jaquet-Droz 1, 2002
Neuchâtel, Switzerland

ABSTRACT

This study contrasts the implantation of 25 µm thick Polydimethylsiloxane (PDMS) membranes with titanium and gold ions at 10 keV and 35 keV for doses from 1×10^{15} at/cm^2 to 2.5×10^{16} at/cm^2 implanted with two different techniques: Filtered Cathodic Vacuum Arc (FCVA) and Low Energy Broad Ion Beam (LEI). The influence of the ion energy, ion type, and implantation tool on the Young's modulus (E), resistivity and structural properties (nanocluster size and location, surface roughness) of PDMS membranes is reported. At a dose of 2.5×10^{16} at/cm^2 and an energy of 10 keV, which for FCVA yields sheet resistance of less than 200 Ω/square, the initial value of E (0.85 MPa) increases much less for FCVA than for LEI. For gold we obtain E of 5 MPa (FCAV) compared to 86 MPa (LEI) and for titanium 0.94 MPa (FCVA) compared to 57 MPa (LEI). Resistivity measurements show better durability for LEI than for FCVA implanted samples and better time stability for gold than for titanium.

INTRODUCTION

Using metal ion implantation (MII) we developed a method that allows micropatterning of very compliant and optically transparent electrodes that can sustain high strains and millions of cycles, and can be used to fabricate buckling type membrane actuators with a vertical displacement of more than 25 % of the membrane's diameter [1]. Starting with a PDMS membrane with a Young's modulus 0.85 MPa, Au ion implantation at 5 keV for doses of order 10^{16} ions/cm^2 results in less than 1 MPa increase in Young's modulus, electrical conductivity below 1 kΩ/ square, and strain of over 175% can be reached while remaining conductive.

In this paper we will present the influence of two different implantation techniques on mechanical, structural and electrical properties of Polydimethylsiloxane (PDMS) implanted with Ti and Au, at 10 keV and 35 keV, and doses between 0.1×10^{16} at/cm^2 and 2.5×10^{16} at/cm^2.

EXPERIMENTAL SETUP

The two implantation instruments, used in this study, were Filtered Cathode Vacuum Arc (FCVA) and Low Energy Broad Beam (LEI) implanter. FCVA creates 600 µs long pulses of plasma with a beam current of 300 µA/cm^2. Pulse rate is of order 1 Hz. The beam is filtered from macroparticles, yielding mostly doubly charged ions. In our case, the beam is then accelerated toward the target with an acceleration potential of 5 kV. The acceleration potential drops during each pulse because of the current from the 17.5×10^{12} at/cm^2 doubly charged ions. LEI delivers a continuous ion current of about 0.5 µA/cm^2. The ion beam is filtered, accelerated and decelerated at the desired energy. Beam energy is stable in time. Averaged over one second, the ion currents are in the same range for both implanters, but during a pulse, the ion current produced by FCVA is much higher than that produced by LEI.

Resistance measurements were performed on 25 μm thick PDMS on Si substrates. Data for surface roughness, dose, mechanical and structural properties were obtained on 25 μm thick PDMS films bonded on patterned silicon chips with 2 mm and 3mm through-holes. The initial Young's modulus was 0.85 MPa. The surface roughness was measured with a Digital Instruments D3100 AFM. The TEM images were obtained with TEM Philips CM, and TEM samples were prepared using a Cryo-ultra-microtome (Leica Ultracut E) at -130°C as described in detail in [2]. A bulge Test setup was used to extract the elasticity of membranes. Once implanted the doses were determined with Rutherford Backscattering Spectroscopy (RBS).

RESULTS

MII creates a nano-composite a few nm thick, located in the top tens of nm of the surface of the elastomer. The particle density and size distribution of clusters depend on the dose, as well as on diffusion. The depth of the implanted layer depends on the ion energy. The thickness of the layers varies according to the ion type and implantation condition, and can be approximately computed for low doses with the TRIM simulation program [3]. Figure 1 shows higher implantation energies lead to larger ion distribution and deeper ion penetration. Gold ions do not penetrate as deeply as titanium, because the stopping cross-section of the ion is proportional to its mass. Hence one may expect gold to form a metallic conducting network at a lower dose and energy than titanium.

Resistance

An important increase of electrical conduction is one of the consequences resulting from MII. The conductivity rises with the dose and depends additionally on ion energy and intrinsic properties of the implanted element (e.g. electronegativity). Microscopically the phenomenon of conduction relates to the size and concentration of the clusters, as well as distance in-between (see figures 3 and 4 for microstructure). As the dose increases, inter-particle spacing decreases as they're getting bigger, till the point where first randomly connected conduction path is formed. This domain is characterized by an abrupt decrease of polymer's resistivity and is referred to as percolation domain. Above this dose the continuous metallic layer is smoothly formed and electrical properties vary only moderately with the concentration, approaching slowly to those of a thin metallic film. The electrical properties of the implanted PDMS are presented in figure 2 and are very well described by the percolation model.

The highest ratio of the electrical conductivity to dose is obtained for gold ions implanted with FCVA at 10 keV (300 Ω/square at 1.75×10^{16} at/cm^2). Gold is a better conductor than titanium and also FCVA produces a narrower distribution of ions due to the drop of the acceleration potential: during each pulse, the gold is initially implanted, and the finally deposited as the acceleration voltage drops (see TEM cross-sections in figure 3 and 4). The slowest decrease of resistance with the dose was measured on the 35 keV titanium samples, as expected from the TRIM simulation.

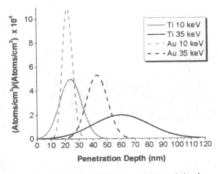

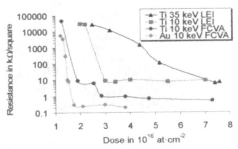

Figure 1. TRIM simulation of Ti and Au ion penetration depth into PDMS versus ion energy. Lower energy leads to denser distribution of ions closer to the surface.

Figure 2. Surface resistance vs. measured ion dose for different energies and implantation techniques. No data is shown for gold implanted by LEI since sputtering prevents obtained conductive layers.

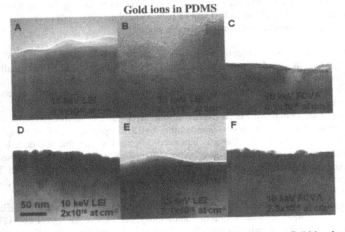

Figure 3. TEM micrographs and cluster morphology of PDMS following Gold ion implantation at 10 and 35 keV using LEI and FCVA. Au implantation leads to round crystalline clusters of diameter up to 30 nm. Doses are measured by RBS. The scale bar is for all 6 photographs. The PDMS is the gray area on the bottom half of each frame. The gold particles are darker, and the carbon grid is white or light gray. The particles depth ranges from 0 nm to roughly 40 nm.

No resistance could be measured for the gold samples implanted with LEI, which is explained by the limited ion concentration. The maximal dose measured by RBS for LEI samples was 2.4×10^{16} at/cm^2. The reason for this is the sputtering of the composite that leads to a steady state condition of the ion concentration.

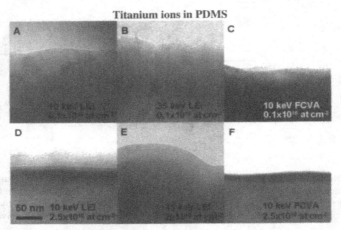

Figure 4. TEM micrographs and cluster morphology of PDMS following Titanium ion implantation at 10 and 35 keV using LEI and FCVA. Ti implantation leads to nm-size amorphous aggregates unlike gold nanoparticles in Figure 3.

Time stability measurements of the conductivity showed that for FCVA implanted samples only the gold samples with doses above 1.75×10^{16} at/cm^2 retained the small electrical resistance. For the higher does, the conductivity remains constant for more than 2 years. At lower doses diffusion of ions and small clusters leads to Ostwald ripening and hence to discontinuity in the conductive network.

Titanium deposited by FCVA forms a thin layer on the surface composed of small clusters that oxidize within few hours. Concerning the LEI implanted titanium samples a small increase of resistance less than 0.1% was observed. Figure 4 shows that the dark layer of the implanted titanium ions is located deeper under the surface and is thus better protected from oxidation. Also stiffening (and probable higher density) of the implanted PDMS layer was observed caused by the radiation induced structural changes. This may prevent the diffusion of the oxygen and the diffusion of the implanted ions.

At high doses the electrical properties of all the samples didn't approach thin film properties, but were showing a saturation of conductivity. For FCVA implantations, AFM observations revealed that, despite the high dose, the metal ions on the surface didn't meld into a continuous film, but stayed clustered. In the case of LEI, the sputtering and the cracking of the surface, due to the increased hardness, prevent the formation of a conductive path.

Microstructure

The microstructure of the nanocomposites is shown in figure 3. Chemical and structural modifications of polymers, and therefore the change of the physical properties, as a function of energy, dose or element have already been studied by other groups [4-8]. The energy transferred from an incoming ion to the polymer matrix can be orders of magnitude higher than typical binding energies of a polymer (15 eV). The bond breaking, producing free radicals, excited species and volatiles, is caused by excitation and ionization of the polymer molecules as well as direct nuclear collisions. If there is a high concentration of primary radicals, then scissioned polymeric bonds react with each other, recombining or initiating cross-linking. Ti implantation with LEI heavily damaged the implanted layer. This leads to cracks in the implanted membrane, which are due to the increased hardness of PDMS. Starting with the surface roughness of 2 nm for a virgin membrane, once implanted the surface buckles into an intricate wavy pattern with root mean square height values of several hundred nanometers (see figure 5).

LEI prepared samples fit quite well with distributions presented in figure 1. FCVA images show that the "implantation" ions are also deposited on the surface, because of the drop of the accelerating potential during each pulse. One also may observe that titanium ions are homogeneously distributed in the matrix forming an amorphous composite, whereas gold clusters into crystalline 30 nm round particles, whose size increases with dose and energy. The reason for this lies in the high chemical reactivity of titanium and the high stability of gold.

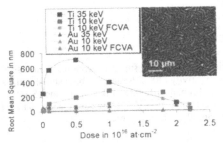

Figure 5. Surface roughness vs. ion dose. Inset: AFM topography for 10 keV FCVA implantation of Au with 2.3×10^{16} at/cm². The PDMS has initial surface roughness of 2 nm. Post irradiation the roughness is due to the different coefficient of thermal expansion between the implanted and non-implanted layer of PDMS. Owing to the very high concentration of ions on the surface, building a continuous metallic layer (see TEM), the roughness decreases again at the higher doses.

Elasticity

The percolation model, described above, elucidates also the changing of elasticity. At the percolation threshold the Young's modulus increases suddenly as a result of the interconnecting metallic clusters (see titanium in figure 6). Nevertheless for a given dose, important differences are observed between LEI and FCVA, as well as between gold and titanium. The low values of the Young's modulus measured on FCVA samples are explained by a partial deposition of the ions thanks to the energy spread from 50 eV to 10 keV. LEI, on the other hand, implants with a mono-energetic beam at 10 keV or 35 keV, rastering and heating the surface of PDMS for several hours. This is a big difference from the short pulsed implantation of a few minutes with FCVA; the influence of the energy on the chemical structure has been shortly summarized above. The Young's modulus for Au rises linearly with dose, contrarily to Ti, whose behavior can be explained with the percolation theory. This is probably due to the diffusion of gold.

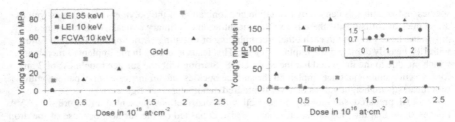

Figure 6. Young's modulus of PDMS membranes after the implantation with Au and Ti. Membranes thickness is 25 µm and the initial value of the Young's modulus is 0.85 MPa. Metal ion implantation has two main effects: formation of metallic clusters in the PDMS, and irradiation-induced chemical modification of the PDMS. A clear percolation threshold is seen for titanium, but not for gold.

CONCLUSIONS

FCVA implantation of gold is best suited for creating conductive electrodes in elastomers since it allows the lowest resistivity (300 Ω/square) to be reached at the smallest dose (1.75×10^{16} at/cm^2) and hence with only limited stiffening of the membrane (5 MPa). This presents the best combination of high electrical conductivity and low compliance that are the main parameters for artificial muscle actuators and flexible electronics.

ACKNOWLEDGMENTS

We thank Dr. M. Doebeli (PSI, Ion Beam Physics) for the RBS measurements and Mr. I. Winkler for his work with LEI at the Center for Application of Ion Beams in Materials Research, Forschungszentrum Dresden-Rossendorf (FZR), Germany. This project was funded by the Swiss National Science Foundation grant #20020-120164, and by the EPFL.

REFERENCES

1. S. Rosset, M. Niklaus, P. Dubois, Herbert R. Shea, Proceedings of SPIE 7287 (2009).
2. M. Niklaus, S. Rosset, P. Dubois, Herbert R. Shea, Scripta Materialia **59**, 893-896 (2008).
3. J. F. Ziegler, *The Stopping & range of ions in matter* (2008). URL www.srim.org.
4. H. Dong, T. Bell, Surface and Coatings Technology **111**, 29-40 (1999).
5. W. Yuguang, Z. Tonghe, L. Andong, Z. Gu, Surface and Coat. Techn. **157**, 262-266 (2002).
6. W. Yuguang, Z. Tonghe, L. Andong, Z. Xu, Z. Gu, Vacuum **69**, 461-466 (2003).
7. W. Yuguang, Z. Tonghe, Z.Gu, Z. Huixing, Z.Xiaoji, Surf.&Coat.Tech. **148**, 221-225 (2001).
8. P. K. Chu, J. Y. Chen, L. P. Wang, N. Huang, Materials Science & Engeneering, Reports **36**, 143-206, (2002).

Mater. Res. Soc. Symp. Proc. Vol. 1188 © 2009 Materials Research Society   1188-LL05-10

Metallic Hollow Sphere Structures Manufacturing Process

C. Davoine, A.Götzfried, S. Mercier, F. Popoff, A. Rafray, M. Thomas, V. Marcadon
ONERA, BP72 - 29, avenue de la Division Leclerc, 92322 Châtillon, France

ABSTRACT

This paper focuses on manufacturing process of regular Metallic Hollow Sphere Structures (MHSS) through brazing technique. As a large stress level is generally confined into the necks formed by brazed spheres, the influence of the filler material on mechanical behavior of cellular metal has been studied. The microstructures of joints resulting from nickel hollow spheres brazing with different commercial fillers "MBF 30" and "MBF 1006" were compared by Scanning Electron Microscopy (SEM) and microhardness testing. These studies revealed a wide boron diffusion into nickel shells through grain boundaries for "MBF 30" brazing, with the formation of borides in a fine brittle eutectic structure. Conversely it was observed that the eutectic structure concentrates at the necks for "MBF 1006" and can be completely eliminated by diffusion-brazing, despite of the shells thinness. The uniaxial compressive tests of HSP specimens have shown two different strain mechanisms depending on brazing process.

INTRODUCTION

Reducing material weight is a permanent challenge for industry, especially for aerospace applications. With this objective, innovative lightweight construction materials were devised in recent years, and cellular metallic materials seem to be a very promising material class [1]. Moreover, cellular metal structures has been gaining interest for multi-functional applications where acoustic or mechanical damping and high temperature application is required. Compared to porous metal which contains a multitude of pores, a cellular metal is a material divided into distinct cells [2]. Honeycombs or MHSS belong to this class of materials. In this paper, we focused on MHSS, but the processes are also studied to devise some other kinds of cellular metal, especially regular array of metallic tubes. The main advantage of this class of material, beside the regularity of its structure which should ensure a good reproducibility of mechanical properties, is to be used in a material-by-design approach [3]. Indeed the thickness and diameter of the cells, the nature of the constitutive metal, the pattern for the piling up can be varied. This allows to design the cellular structure in order to reach the target properties imposed on the material by design requirements.

Hollow spheres can be produced through galvanic methods or through a powder metallurgy based manufacturing process. In both cases, EPS (expanded poly styrol) spheres are used to obtain the spherical shape before being eliminated through pyrolyse or through chemical dissolution. Various joining technologies such as brazing and adhering can be used to assemble the single hollow spheres together. An advantage of brazing hollow spheres is to create a neck, whose mechanical behavior and morphology could be used as a further design parameter for the optimisation of the structure's mechanical properties for specific applications. Here, the joining of single hollow spheres by brazing of nickel-based foils has been investigated. Nickel-based fillers metals usually contain several "melting point depressant" which can form eutectic structures that are extremely hard and brittle in the brazed joint [4]. Brazing by allowing a sufficient amount of time for a complete isothermal solidification to take place at the brazing temperature could prevent the formation of intermetallics during cooling [5]. In case of cells such as hollow spheres, the amount of base metal is finite, and the feasibility of diffusion-brazing is determined by the limit of solubility of the melting point depressants.

EXPERIMENTAL WORK

Small regular samples of nickel hollow spheres provided by the French firm Ateca (2mm in diameter with a shell thickness of 150μm) have been manufactured using commercially available brazing foils. The foils have been placed between planes of spheres, as shown in figure 1.a), in a graphite mould which ensures the contact. During the thermal treatment, the brazing foils melt and the liquid wets the spheres surface. As the liquid concentrates at the contact points by capillarity, both vertical and horizontal necks are created (figure 1.c)). After cooling, the SC-like MHSS samples exhibit a suitable regularity, as shown in figure 1.b).

nickel-based
brazing foil

nickel hollow
spheres

Figure 1: a) Stackings of hollow spheres and brazing foils b) SC-like MHSS c) Neck between two brazed hollow spheres.

Two kinds of amorphous nickel-based brazing foils provided by MetGlas have been used: "MBF 30" and "MBF 1006". Their composition, melting range temperature and thickness are given in table 1. In both cases, the samples were heat treated to reach a temperature slightly higher than the liquidus but lower than the melting temperature of the pure Ni, to avoid the fusion of the spheres. The heating speed has been limited to 10°C/min to prevent the explosion of hollow spheres caused by the thermal dilatation of the entrapped gas. To finish, the samples are cooled to room temperature following the natural cooling of the furnace. For "MBF 30" the heat treatment has been carried out in neutral atmosphere (argon), and for "MBF 1006" it has been performed under vacuum. Two pure nickel foils were brazed at the bottom and at the top of each cubes to consolidate the specimens.

	Nominal composition (%wt)				Melting T	Thickness
	Pd	B	Si	C	Ni	(°C)	(μm)
MBF 1006	41.2	-	8.8	-	Bal	714-938	40
MBF 30	-	3.2	4.5	0.06	Bal	984-1054	38.1

Table 1: Characteristics of brazing foils. (MetGlas source)

RESULTS AND DISCUSSION

Microstructures

The figure 2 a) showing a neck between two hollow spheres illustrates the SEM observations of a cross-sectional cut of "MBF 30"- brazed samples heat treated at 1180°C during 30min. The atomic number based BackScatter Electron (BSE) images showed the microstructure to be consisting of two distinct areas corresponding to the nickel based solid solution (Ni) and eutectic structure. The ternary system suggests that the eutectic stucture is composed by intermetallic borides (Ni_3B) in a nickel rich phase. Indeed, the surface of the spheres revealed the crystalline aspect of (Ni) grains and borides localized at grain boundaries. In case of "MBF 30" brazing, boron diffuses extensively out of the neck into nickel shells during the heat treatment. Instead of being localized at the necks, the liquid phase propagates through the shells which are 150μm thick.

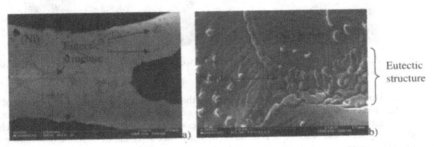

Eutectic structure

Figure 2: SEM observations of brazed nickel hollow spheres with "MBF 30" foils. Picture a) shows a neck between two hollow spheres, while picture b) shows the intermetallic borides at grain boundaries at sphere's surfaces.

Although the boron content in (Ni) grains cannot be measured by EDS compositional analysis, we assume that this microstructure is formed as a result of diffusion of boron exceeding the solubility limit at the bonding temperature. Schobel reported a value of 0.3 at.% for the maximum solubility of B in (Ni) at the eutectic temperature of 1093 °C of the binary system [6]. Therefore the amount of brazing material needed to create sufficiently large necks is superior to the amount that it is possible to dissolve in the nickel shells. Intermetallic phases are known to be detrimental to base metal mechanical behavior. Microhardness of the microstructure was measured by using a microhardness tester with a load of 0.3N and a loading speed of 0.2g/s. A series of 35 tests were conducted both in the shells and in the necks. The average hardness of the eutectic structure was found to be 5.1GPa, which is much higher than the hardness of the (Ni) grains (1.2 GPa in average). Brazing filler metals should be chosen to minimize the amount of any brittle phases to provide high joint ductility. Although the existence of two intermetallic phases have been reported ($NiPd_2Si$ and $Ni_{18}Pd_7Si_9$) in the Ni-Pd-Si system, the Ni-Pd system is an isomorphous system with a minimum of 1237°C at 45 at.% Pd [7]. Palladium appears thus as a good candidate to avoid intermetallics formation thanks to diffusion-brazing.

The samples made by stacking of nickel hollow spheres and "MBF 1006" brazing foils were heat treated at 1150°C during 1h. SEM-observations are given in figure 3. While hollow spheres samples brazed with "MBF 30" formed an eutectic structure localized at grain boudaries through the shell, the microstructure of all the samples brazed with "MBF 1006" consisted of a continuously distributed centerline solidification constituent localized in the necks. The SEM-observations in BSE-mode and EDS compositional analysis showed this area to be consisting of three distinct areas: nickel based solid solution (Ni), $NiPd_2Si$ grains and the eutectic structure with a high content in palladium.

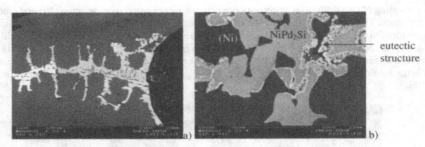

eutectic
structure

Figure 3: SEM observations of brazed necks between hollow spheres brazed with "MBF1006" 1h at 1150°C.

To study the evolution of the microstructure through diffusion-brazing, the influence of brazing holding time (1h, 2h and 16h at 1200°C) on the formation of this centerline eutectic was investigated. It is seen that the eutectic area decreased with holding time. Figure 4.a), which shows the SEM observations in BSE mode of a cross-sectional cut of a sample held during 16h reveals a monophase microstructure. The corresponding EDS compositional analysis performed through a neck confirms the absence of a multiphase microstructure (figure 4.b)). The complete isothermal solidification of the transient liquid phase that temporarily exists at the brazing temperature has thus been taken place, and prevented the formation of brittle deleterious phases during cooling.

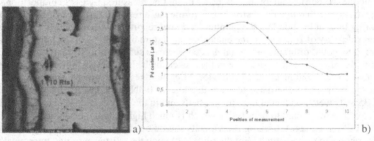

Figure 4: a) SEM observations of brazed necks between hollow spheres with "MBF1006" foils: 16h at 1200°C; b) Palladium content mesurement through the neck.

<u>Compressive tests</u>

An experimental investigation of the quasi-static uniaxial compressive loading of the 4x4x4 cubic samples has been carried out. The tests were conducted on a servohydraulic load frame (100KN) under displacement control. The load was recorded at a prescribed displacement rate of 1 mm/min. Thanks to the value of the section area and the height of the sample, the force vs. displacement curves have been converted into global stress vs. strain curves.

Samples "MBF 30"-brazed held at 1180°C during 1h ,"MBF 1006"-brazed held at 1200°C during 1h and the same held during 16h called "MBF 1006 H" were tested until complete crumbling. The stress vs. strain curves given by the figure 5 a) highlight two different mechanical behaviors. From a global strain of 10%, the "MBF 30" and "MBF 1006" brazed sample crumbled suddenly, while the "MBF 1006 H" did not exhibit any macroscopic damages. This crumbling of "MBF 30" brazed sample could be explained by the microstructure resulting from brazing process described previously. Indeed the presence of a hard eutectic structure in the nickel shells caused by the formation of borides at grain boundaries results in a mechanical heterogeneous structure. During loading, the eutectic structure probably served as a preferred path for crack initiation and propagation.

Conversely the maximal global strain of "MBF 1006 H" brazed samples reached more than 70% without generating any macroscopic damages, as shown in figure 5 b). The pictures of crushed samples indicate that their global strain is associated with the plastic crushing of hollow spheres. The shape of hollow spheres changed during the compressive test from a spheroid to an elliptical morphology. The consequence of this geometrical hardening is an increased strength and Young's modulus for the MHSS with increasing global straining. The curves do not exhibit any plateau stress, because plastic deformation of the necks occurs from the very beginning. Moreover the measurements performed on set of seven identical samples present an excellent reproducibility.

The joining of nickel hollow spheres through diffusion-brazing seems to provide ductility to the cellular material. Such a large global strain could be interesting for impact-energy-absorbing devices. The main result of this first experiments is that the brazing filler has a noteworthy influence on mechanical behavior of cellular materials manufactured through the brazing of individual cells.

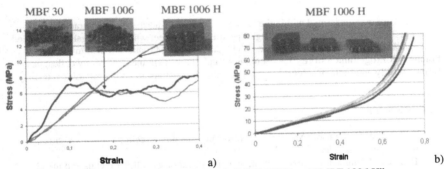

Figure 5: Strain *vs.* stress curves of a) "MBF 30", "MBF 1006" and "MBF 1006 H" homogenized, b) the seven homogenized "MBF 1006 H" samples.

CONCLUSIONS AND PERSPECTIVES

This study showed the influence of the brazing material associated with brazing parameters on global strain of regular MHSS under compressive loading. Two different strain mechanisms have been observed. The first one, resulting from a classical brazing process of "MBF 30" foils, causes the rupture of the shells and the crumbling of the sample. The samples exhibit such a ductility, that they can be crushed until a large deformation without macroscopic damage. The second strain mechanism, resulting from a diffusion-brazing bonding of "MBF 1006" foils, corresponds to the progressive plastic crushing of hollow spheres.

Therefore, brittle behavior of the cellular structure due to the presence of intermetallics into the shells can be prevented by homogenization of the microstructure through diffusion-brazing. Indeed efforts to reduce the formation of intermetallics phases through a significant holding at brazing temperature, improves the ductility of cellular samples which can reach large global strain without crumbling. But the complete isothermal solidification of the transient liquid phase depends on the maximal solubility of elements in the base metal. While diffusion-brazing has evolved into a successful joining technique for nickel hollow spheres with "MBF 1006" foils, the microstructure resulting from "MBF 30" foils brazing cannot be homogenized because of the thinness of the nickel shells. Indeed the minimal amount of boron needed for amorphability of the brazing foils can not be completely dissolved, and results in the formation of an eutectic structure, whose average hardness measured by microhardness tester was found to be much higher than the hardness of nickel solid solution. Quasi-static compression tests have shown that the confinement of the eutectic structure at grain boundaries through the shells had a detrimental effect on mechanical behavior.

The brazing of nickel hollow spheres provided regular "model" samples. At the same time, the manufacturing process of stainless steel and superalloy hollow spheres was studied. The design of another kind of cellular material, composed of tubes instead of hollow spheres is also in progress. Future work will focus on diffusion-brazing of nickel-based brazing material into another base metal than nickel.

ACKNOWLEDGMENTS

This work was performed within the "Aerodynamic and Thermal Load Interactions with Lightweight Advanced Materials for High Speed Flight" project which investigates high-speed transport. ATLLAS, coordinated by ESA-ESTEC, is supported by the EU within the 6th Framework Programme Priority 1.4, Aeronautic and Space. I am indebted to my colleagues Bruno Passilly and François-Henri Leroy for fruitful discussions about microhardness testing.

REFERENCES

1. A. Evans, J.W Hutchinson, M.F.Ashby, "Multifunctionality of cellular metal systems", *Progress in materials science*, **43**, 171-221 (1998)
2. J. Banhart, "Manufacture, characterisation and application of cellular metals and metal foams", *Progress in Materials Science*, **46**, 559-632 (2001)
3. A. Fallet, P. Lhuissier, L. Salvo ,Y. Bréchet, "Mechanical Behaviour of Metallic Hollow Spheres Foam", *Advanced engineering materials* **10,** No. 9, 858-862 (2008)
4. W.D. Zhuang, T.W. Eagar, "High Temperature Brazing by liquid Infiltration", *Proceeding of the 26ᵗʰ International Conference on Brazing and Soldering*, 526-531 (1997)
5. M. C. Chaturvedi, O. A. Ojo and N. L. Richards, "Diffusion Brazing of Cast Inconel 738 Superalloy", *Advances in Technology of Materials and Materials Processing*, **1**, 1-12 (2005)
6. J.D. Schobel and H.H. Stadelmaier, *Z. Metallkd.*, **56** (12), 856-859 (1965)
7. K. P. Gupta, "The Ni–Pd–Si (Nickel-Palladium-Silicon) System", *Journal of Phase Equilibria and Diffusion*, **27**, No. 4 (2006)

Mater. Res. Soc. Symp. Proc. Vol. 1188 © 2009 Materials Research Society   1188-LL04-05

Bio-Based Polyurethane-Clay Nanocomposite Foams: Syntheses and Properties

Min Liu[1], Zoran S. Petrovic[1] and Yijin Xu[1],*
[1]Kansas Polymer Research Center, Pittsburg State University, Pittsburg, KS 66762, U.S.A.
* To whom all the correspondences should be addressed.

ABSTRACT

Starting from a bio-based polyol through modification of soybean oil, BIOH™ X-210, two series of bio-based polyurethanes-clay nanocomposite foams have been prepared. The effects of organically-modified clay types and loadings on foam morphology, cell structure, and the mechanical and thermal properties of these bio-based polyurethanes-clay nanocomposite foams have been studied with optical microscopy, compression test, thermal conductivity, DMA and TGA characterization. Density of nanocomposite foams decreases with the increase of clay loadings, while reduced 10% compressive stress and yield stress keep constant up to 2.5% clay loading in polyol. The friability of rigid polyurethane-clay nanocomposite foams is high than that of foam without clay, and the friability for nanofoams from Cloisite® 10A is higher than that from 30B at the same clay loadings. The incorporation of clay nanoplatelets decreases the cell size in nanocomposite foams, meanwhile increases the cell density; which would be helpful in terms of improving thermal insulation properties. All the nanocomposite foams were characterized by increased closed cell content compared with the control foam from X-210 without clay, suggesting the potential to improve thermal insulation of rigid polyurethane foams by utilizing organically modified clay. Incorporation of clay into rigid polyurethane foams results in the increase in glass transition temperature: the Tg increased from 186 to 197 to 204 °C when 30B concentration in X-210 increased from 0 to 0.5 to 2.5%, respectively. Even though the thermal conductivity of nanocomposite foams from 30B is lower than or equal to that of rigid polyurethane control foam from X-210, thermal conductivity of nanocomposite foams from 10A is higher than that of control at all 10A concentrations. The reason for this abnormal phenomenon is not clear at this moment; investigation on this is on progress.

INTRODUCTION

The utilization of natural products such as plant oils and natural fats has attracted great attentions in both scientific and industrial areas in recent years, due to both the energy and the environmental considerations. These starting materials are sustainable, renewable, and most importantly, biodegradable. Rigid polyurethane foams have been widely used for insulation purposes due to their better thermal properties than those of thermoplastic foams and other common insulation materials. It has been confirmed that polymer-clay nanocomposites show improved thermo-mechanical properties, better scratch resistance, improved gas permeation property, higher glass transition and stiffness, and better flame retardancy. Starting from a bio-based polyol through modification of soybean oil, a series of bio-based polyurethanes and their clay nanocomposite foams have been prepared. The effects of organically-modified clay types and loadings on foam morphology, cell structure, and the mechanical and thermal properties of these bio-based polyurethanes-clay nanocomposite foams have been studied with optical microscopy, compression test, DMA and TGA characterization. The effect of clay types and loadings on the friability, closed cell content, and thermal conductivity of the bio-based nanocomposite foams were also characterized and discussed.

EXPERIMENTAL DETAILS

Materials. Soy polyol X-210 is a BIOH™ product from Cargill with a OH# of 225 mg KOH/g and OH functionality of 4.4; Cloisite®10A and 30B were kindly provided by Southern Clay Products Inc.; DABCO® DC198 Surfactant, DABCO® LK® 433 Surfactant, DABCO® T-12 Catalyst were obtained from Air Products and Chemicals, Inc.. Rubinate® M (Polymeric Methylene Diphenyl Diisocyanate) was kindly provided by Huntsman Americas. NIAX® Catalyst A-1 was gift from OSi Specialties, Inc.. DIUF water and Glycerol (ReagentPlus®, 99% GC) were purchased from Fisher Scientific. All the reagents were used as received.

Instrumentation. Viscosity of Soy polyol-nanoclay mixtures was measured on a AR 2000EX Rheometer, TA Instruments Ltd.. Mechanical properties were measured on QTest, Sintech according to ASTM D1621-94. Closed cell content was obtained with Quantachrome Untrapycnometer 1000 based on the method ASTM D6226. DMA 2980 Dynamic Mechanical Analyzer and TGA 2050 Thermogravimetric Analyzer (TA instruments, Inc.) were used to characterize the thermal properties of polyurethane-clay nanocomposite foams. The friability was measured on home-made equipment according to ASTM C421-88. The thermal conductivity of foams was measured with Heat Flow Meter (Model LaserComp Fox 200, LaserComp's Inc.) with LaserComp's "WinTherm32" software according to ASTM C518-91. Optical Microscope-polarized, (Model: Rolam 312, manufacturer: LOMO) equipped with a Cannon EOS Digital 400D camera was used to get the images for cell structure, the images were analyzed with Software Image Tool to obtain information on cell size and cell density.

Preparation of Soy polyol-Nanoclay Mixtures. Soy polyol BIOH™ X-210 in a one-necked round-bottomed flask was heated to 80°C with magnetic stirring, and then prescribed amount of nanoclay was slowly added into polyol with a solid powder addition funnel. The mixture was further stirred for 1h, after which the flask was put in a sonicator preheated at 60°C and sonicated for 10 minutes. All the polyol-clay mixtures looked homogenous and no settlement of clay was observed after standing for several weeks at room temperature. Polyol-clay mixtures with five clay concentrations (0%, 0.5%, 1.5%, 2.5%, and 5%) from two clays (Cloisite® 10A and 30B) were prepared.

Preparation of Rigid Polyurethane Foams. Soy polyol X-210 or X-210-clay mixture, catalysts, water, and surfactant were weighed into a 500-mL disposable plastics cup and mixed with a mechanical stirrer at 3000 rpm for 25s; then prescribed amount of polyisocyanate Rubinate® M was added and mixed for 12s; after which the mixture was quickly poured into a 5-L paper bucket. All the PU foams were cured for one week at room temperature before characterizations.

DISCUSSION

Figure 1 illustrates the chemical structures of BIOH™ X-210 soy polyol, and quaternary ammonium cations used in Cloisite®10A and 30B; T and HT stand for tallow and hydrogenated tallow fatty chain, respectively.

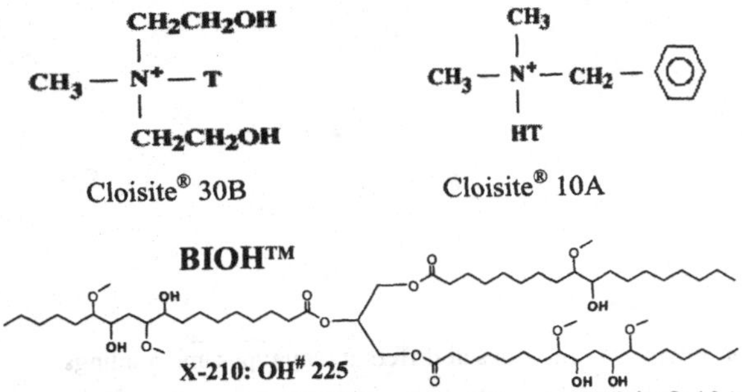

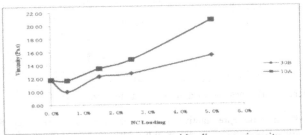

Figure 1. Structures for X-210 and modifiers in Cloisite® 10A and 30B.

Figure 2 shows the effect of clay types and loadings on polyol-clay mixture viscosity. Even though the viscosity of mixtures from X-210 and Cloisite® 10A is higher than that of pure X-210 beyond clay concentration of 0.5%; the viscosities from X-210 and 30B mixtures were equal to or lower than that of pristine X-210 up to 30B loading of 2.5%. The viscosity of mixtures from 10A is always higher than that from 30B at the same clay concentration.

Figure 2. Effect of clay types and loadings on viscosity

The cream and rise times for different X-210/clay mixtures are summarized in the **Figure 3**. The cream time for clay mixtures (24s for 5.0% 30B and 21s for 5.0% 10A) is shorter than that for control X-210 (28s for pure X210) and it decreased with the increased clay loading. The rise time for mixtures with clay (130s for 5.0% 30B and 97s for 5.0% 10A) were also shorter compared to that for control X-210 (168s for pure X210). Both cream and rise times for the mixtures with 10A were shorter than those for mixtures with 30B, indicating higher catalytic effect of 10A on urethane formation.

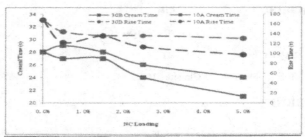

Figure 3. Cream and Rise time for polyol-clay mixtures at different clay types and loadings.

The densities of rigid PU nanocomposite foams decreased with increased 30B and 10A clay concentration, which are lower than that of X210 control foam as shown in **Figure 4**, except for the foam with 0.25% of 10A.

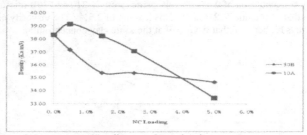

Figure 4. Effects of clay types and loadings on foam density.

Figure 5 collects some representative images of cell structures for different foams at different clay loadings by optical microscopy. It is clearly seen that the cell size is reduced a lot from pristine X-210 to the foams from polyol with 0.5% of organically-modified clay, beyond which it almost keeps unaltered with respect to clay concentration. **Figure 6** shows the cell size and cell density [3] for nanofoams from 30B and 10A at different clay loadings. The cell size decreased from 0.16 mm to 0.108 mm and 0.102 mm for 0.5% 30B loading and 0.5% 10A loading, respectively.

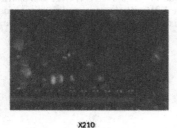

X210

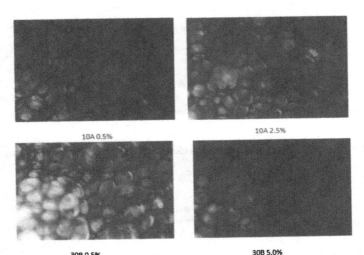

10A 0.5%

10A 2.5%

30B 0.5%

30B 5.0%

Figure 5. Optical microscopy images of the nanofoams at different clay loadings. All figures have the same magnification.

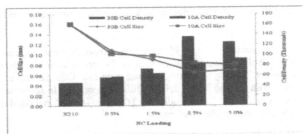

Figure 6. Cell size and cell density of PU foams and nanofoams at different clay types and loadings.

Figure 7 depicts the effect of clay loadings on Closed Cell Content (CCC) of bio-based nanofoams. On contrary to some reported results, the addition of clay resulted in increased closed cell content in our systems, this is probably due to the better miscibility between our bio-based polyol and the modifiers in 30B and 10A. Higher CCC usually leads to better thermal insulation, which is a desired property.

Figure 8 summarizes the 10% compressive stresses and Yield stresses for nanofoams and control. It is known that mechanical properties of cellular materials depend mainly on their density, thus the 10% compressive stress and Yield stress of all foams have been normalized against their densities. For foams from both 10A and 30B nanoclays, their compressive strengths remain unaltered up to 2.5% of clay loadings; after which, the value decreased sharply in the case of 10A. Yield stresses for foams with both fillers showed the same trends.

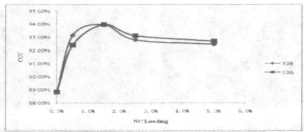

Figure 7. Effects of clay types and loadings on closed cell content (CCC) of nanofoams.

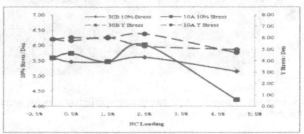

Figure 8. Effects of clay types and loadings on 10% compressive stress and yield stress of rigid PU nanocomposite foams.

Glass transition temperatures of rigid PU-clay nanocomposite foams increased with the addition of clay and with clay concentrations, which can be clearly seen from the representative results shown in **Figure 9**. The Tg of foams increased from 186 °C for control to 198 °C and 204 °C for nanofoams with 0.5 and 2.5% Cloisite® 30B, respectively. In accordance with this trend, the thermal stability of nanofoams studied by TGA also showed increased decomposition peak temperatures with increased clay loadings.

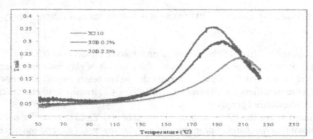

Figure 9. DMTA results of bio-based PU control and PU-clay nanocomposite foams.

There is no linear relationship between the friability and nanoclay concentration, as shown in **Figure 10**. Generally, the addition of nanoclay resulted in higher friability in foams, the effect of 10A on friability increase is much more significant that that of 30B, which is probably due to the presence of hydroxyl groups in 30B which leads to the incorporation of 30B into polyurethane networks, while 10A only acts as physical additive.

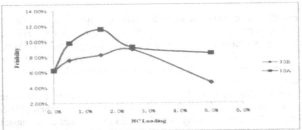

Figure 10. Effects of clay types and loadings on the friability of rigid PU nanocomposite foams.

Figure 11 shows the effect of clay types and clay loadings on thermal conductivity of polyurethane-clay nanocomposite foams. It is well-known that in rigid PU foam lower thermal conductivity is achieved with smaller cell size. From **Figure 6** one can find that the cell size is reduced significantly with the addition of only 0.5% of organically modified clay in X-210, after which it keeps unaltered up to the clay concentration of about 5%. However, as shown in **Figure 11**, as the concentration of clay increased, we didn't see significant change in thermal conductivity for the nanofoams with 30B; and what's more interesting is that the thermal conductivity of nanofoams from Cloisite®10A is higher than, rather than lower than that of control foam. The reason for such abnormal phenomena is not clear at this moment and is under investigation.

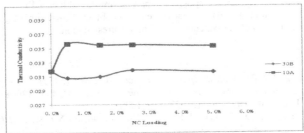

Figure 11. Effects of clay type and loadings on thermal conductivity of nanofoams.

CONCLUSIONS

From above-discussions we come to the following conclusions:
1)   The mixtures from X-210 and organically-modified clay are stable for several weeks in our research, which is different from mixtures from petro-based polyol and organically-modified clay, in which settling down of clay happens within 24 hours. The stability of bio-based polyol

and organically-modified clay mixture is due to the miscibility of fatty chain in organic modifiers for clay and the characteristic fatty chain structures in bio-based polyol.

2)   Both cream time and rise time became shorter with increase of clay concentrations in the polyol-clay mixtures, indicating the catalytic effect from quaternary ammonium species on modified clays.

3)   The incorporation of organically-modified clay in polyol results in decreased density of polyurethane-clay nanocomposite foams; while the 10% compressive stress and yield stress normalized against density almost kept unchanged up to a clay loading of 2.5% in polyol.

4)   With the increase of clay loading, cell size of nanofoams decreases while the corresponding cell density increases. Smaller cell size provides the possibility to obtain better thermal insulation property of the nanofoams.

5)   Incorporation of clay into rigid polyurethane foams results in the increase in glass transition temperature: the Tg increased from 186 to 197 to 204 °C when 30B concentration in X-210 from 0 to 0.5 to 2.5%, respectively; which would provide better dimensional stability for the resulted nanofoams; in accordance with this, thermal stability of corresponding nanofoams tested on TGA showed systematic increase in decomposition peak temperatures.

6)   Friability of nanocomposite foams is higher than that of rigid polyurethane foam from pure X-210; the value for foams from 30B is lower than that of foams from 10A at the same clay concentration, this is probably due to the presence of hydroxyl groups in 30B which leads to the incorporation of 30B into polyurethane networks, while 10A only acts as physical additive.

7)   Even though the thermal conductivity of nanocomposite foams from 30B is lower than or equal to that of rigid polyurethane foam from pristine X-210, thermal conductivity of nanocomposite foams from 10A is higher than that of rigid polyurethane foam from pristine X-210 at all 10A concentrations. The reason for this abnormal phenomenon is not clear at this moment; investigation on this is on progress.

8)   Based on the preliminary results discussed above, together with the improved flame-retardance and barrier properties in polymer-clay nanocomposites, it would be reliable to say that bio-based polyurethane-clay nanocomposite foams with improved thermal insulation properties, dimensional stability, and better flame-retardance can be obtained by optimizing the formulation and processing parameters of the systems studied in this presentation.

REFERENCES

1. A. Guo, I. Javni, Z.S. Petrovic. *J. Appl. Polym. Sci.* 77(2), 467 (2000).
2. S. Pavlidou, C.D. Papaspyrides. *Progress in Polymer Science* 33, 1119 (2008).
3. X. Cao, L.J. Lee, et al. *Polymer* 46(3): 775-783 (2005).
4. L.J. Lee, C. Zeng, X. Cao, X. Han, J. Shen, G. Xu. *Composites Science and Technology*, 65, 2344 (2005).
5. T. Widya and C. W. Macosko. *J. Macromol. Sci. Part B: Physics*, 44, 897 (2005).
6. G. Harikrishnan, T. Umasankar Patro, and D.V. Khakhar. *Ind. Eng. Chem. Res.*, 45, 7126 (2006).
7. M. Thirumal, Dipak Khastgir, N.K. Singha, B.S. Manjunath, and Y.P. Naik. *Cellular Polymers.* 26, 245 (2007).

Mater. Res. Soc. Symp. Proc. Vol. 1188 © 2009 Materials Research Society   1188-LL06-07

Anisotropic Clay Aerogel Composite Materials

Matthew D. Gawryla, David A. Schiraldi
Department of Macromolecular Science and Engineering, Case Western Reserve University,
2100 Adelbert Road, Cleveland, OH 44106

Abstract

Clay aerogel composites have been around for over 50 years but still they represent a relatively under studied class of materials. Clay aerogel composites have been made in our labs that have low densities, $0.05-0.1 g/cm^3$, provide good thermal insulation, $k \approx 0.02W/mK$, and are created through an environmentally benign process. The mechanical properties of the composites resemble those of typical foamed polymers such as expanded polystyrene and polyurethane, with compressive moduli ranging from 0.5MPa to 40MPa depending on composition. Aqueous solutions of clay and polymer are frozen in cylindrical molds and freeze-dried to create these foam-like materials. Typically there is no particular orientation to the often layered structure that results, however if frozen in a unidirectional manner, anisotropic materials can be made. In this paper we will discuss the effects of molecular weight on mechanical properties of various composites as well as discussing the orientated layered structure within the anisotropic materials.

Keywords:

Aerogel, Anisotropy, Clay, Composite

Introduction

Clay aerogels and clay aerogel/polymer composites are relatively new members of a family of low density materials created through ice templating [1-3]. The use of water soluble polymers and sodium montmorillonite (Na^+-MMT) can create materials with mechanical properties comparable to typical foamed polymers [4-7]. The process of ice templating proceeds as depicted in **Figure 1**, with any non-water material being forced between the growing ice crystals. This occurs due to the crystal wanting to retain as pure a structure as possible and excluding all non-water molecules/matter from the crystal. Once the sample is frozen, the water is sublimed away via freeze-drying, leaving only the templated solid material behind. If the solid material interacts with itself, or there is a reinforcing polymer present, the three dimensional structure will usually be self supporting. *Gutierrez* has shown that under certain processing conditions, the molecular weight of the polymer plays a significant role in the final structure [8].

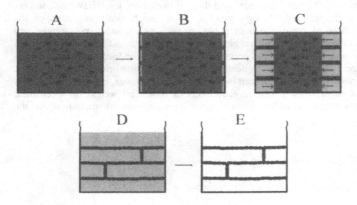

Figure 1. Ice templating process for randomly nucleated samples - (A) Stable gel/solution (B) Nucleation at the edge of the vial (C) Ice growth toward the center of the vial (D) Frozen material (E) After sublimation

A material is described as anisotropic if the properties differ with respect to orientation of the sample. Wood is a very good example of an anisotropic material, having high strength in the in grain direction, and lesser properties in the other two directions [9]. Anisotropy is useful in creating a range of products, from piezoelectrics to energy absorbing panels [10]. Using a vertical freezing method it is possible to create highly anisotropic materials which can be used to determine the mechanical properties of a single domain for use in modeling the randomly nucleated ice templated structures. In addition to verification and assistance in modeling, these anisotropic materials can be used as a core material for composites which need to bend but retain their rigid nature. An example of this could be an airplane wing which needs to be rigid and retain the airfoil cross section in the Z-direction of the wing but still flex in the X-Y plain of the entire structure.

Sodium montmorillonite (Na⁺-MMT), PGW grade, (Nanocor Inc.), and 13-24KD, 31-50KD, 85-124KD and 146-186KD poly(vinyl alcohol) (PVOH), (Sigma Aldrich) were used as received. Clay dispersions were created using a Waring model MC2 mini laboratory blender with deionized water prepared using a Barnstead RoPure reverse osmosis system. Samples were freeze-dried using a Virtus Advantage EL-85 freeze-drier at an ultimate pressure of 5 μbar and a temperature of 25°C.

Figure 2. Custom mold for creating anisotropic materials (A) polypropylene sides (B) aluminum base (C) freeze-drier shelf, -70°C.

Anisotropic materials were prepared by mixing 5 wt.% PVOH solutions with the 10 wt.% clay gels and freezing them in a vertical manner using the mold shown in **Figure 2**, followed by drying. Cubic samples were cut from the inner portions of the vertically oriented samples so as to eliminate any effect the edge may have had on freezing. The cubes were tested in two orientations (**Figure 3**) in order to investigate the anisotropic nature of samples prepared in this manner.

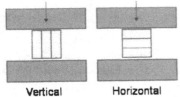

Vertical   Horizontal

Figure 3. Orientation of layers for testing of mechanical anisotropy

Results/Discussion

The mold in **Figure 2** was designed so that thermal energy is transferred in the vertical direction and only a minimal amount is removed through the sides. Polymeric sides were chosen as an easy to use material with a relatively low thermal conductivity as compared to that of aluminum or water. As the thickness of the solution to be frozen increases, the ability to remove thermal energy decreases over the freezing period. The slowing of the thermal energy removal leads to a decrease in the rate of ice growth through the liquid material. Decreasing the rate of freezing allows the smaller crystals to coalesce into larger domains which results in thicker layers with wider spacing in between each layer. In order to keep a consistent, uniform material, samples were limited to approximately 1 cm in thickness.

Figure 4 shows the micro-structure of the materials changing with increasing molecular weight from a nicely ordered lamellar structure to a disordered structure. This change is likely due to the association of the larger molecular weight polymer chains with more than one clay platelet, creating a stronger gel network. The stronger network prevents the ice crystals from growing continuously through the material during freezing resulting in the disordered structures observed for the materials made with the two higher molecular weight polymers.

Figure 4. Ice templated composites (A) 5wt.% Na$^+$-MMT/2.5wt.% 13-24KD PVOH (B) 5wt.% Na$^+$-MMT/2.5wt.% 146-186KD PVOH

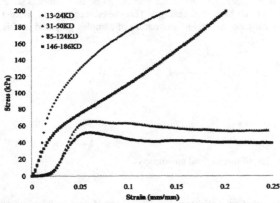

Figure 5. Stress strain curve of vertically compressed anisotropic samples

The two lower molecular weight samples, compressed in the vertical direction, failed by abrupt layer buckling (**Figure 5**). The distinct yield point in these two samples at approximately 5% strain indicates the irreversible deformation of the layers, after which the properties remain level or decrease slightly. The small induction period seen prior to the linear, elastic portion of the curve is due to sample manufacturing. It has been found that as long as the magnitude of the induction period is less than the linear region of the curve it has very little effect on the measured elastic modulus. Yield strain can be measured by plotting a line through the linear region, finding its x-intercept and subtracting that from the measured strain value. The more disordered structures, 85-124KD and 146-186KD, lack the long layers and therefore do not fail in the

106

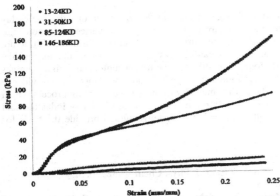

Figure 6. Stress Strain curve of horizontally compressed anisotropic samples

same manner. Compression of these higher molecular weight samples can be described as densification rather than abrupt failure due to the continual increase in the force required to compress them.

The low molecular weight horizontal samples showed a gradual yielding, similar to what is seen in elastomeric materials and soft foams (**Figure 6**). The layers provide a flexible structure in this dimension, showing very little hysteresis upon multiple compression cycles. Quantifying the mechanical anisotropy is done by comparison of the modulus in the vertical direction to that of the horizontal. The samples exhibit significant anisotropic behavior at low molecular weights, a 24x and 19x difference from vertical to horizontal for the 13-24KD and 31-50KD samples respectively, and only slight differences in the higher molecular weight at 3x for the 85-124KD and 2x for the 146-186KD samples (**Figure 7**).

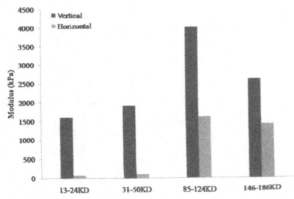

Figure 7. Graphical representation of the anisotropy seen in vertically frozen samples

Conclusions

The directional freezing of clay/PVOH gels results in the formation of anisotropic clay aerogel/PVOH composites. The internal structure of these composites varies with molecular weight from highly lamellar to a random/disordered structure. The change in internal structure resulting from the different molecular weights causes drastic differences in the mechanical anisotropy of the samples. This study found that although the higher molecular weight samples appear isotropic in the electron microscope, the samples still retain some mechanical evidence from the direction they were frozen. The environmentally friendly ice templating process which has been described can be used to create anisotropic materials for use in many industrial and commercial applications and the particular clay/polymer systems studied provide the basis for further studies to come.

References

[1] S. S. Kistler, Nature 1931, 127, 741; S. S. Kistler, J. Phys. Chem. 1932, 36, 52

[2] H. Van Olphen, Clay Miner. 1967, 15, 423-435

[3] L. S. Somlai, S. A. Bandi, L. J. Mathias, D. A. Schiraldi, AICHE J. 2006, 52, 1162

[4] M. D. Gawryla, M. Nezamzadeh, D. A. Schiraldi, Green Chemistry, 2008, 10, 1078

[5] E. Arndt, M. D. Gawryla, D. A. Schiraldi, J. Mater. Chem. 2007, 17, 3525

[6] K. A. Finlay, M. D. Gawryla, D. A. Schiraldi, J. Indust. Eng. Chem. Res. 2008, 47, 615

[7] M. Gutierrez, M. L. Ferrer, F. del Monte, Chem. Mater., 20, pg 634, 2008.

[8] M. C. Gutierrez, Z. Y. Garcia-Carvajal, M. J. Hortiguela, L. Yuste, F. Rojo, M. L. Ferrer, F. del Monte, Adv. Funct. Mater. 2007, 17, 3505-3515

[9] K. E. Easterling, R. Harrysson, L. J. Gibson, M. F. Ashby; "On the mechanics of balsa and other woods" Proc. R. Soc. Lond. A, 383, pg 31, 1982.

[10] L. J. Gibson, M. F. Ashby; "Cellular Solids – Structure and Properties 2nd Ed." Cambridge Univ. Press, 152, 1997.

Towards Structures

Mater. Res. Soc. Symp. Proc. Vol. 1188 © 2009 Materials Research Society   1188-LL05-01

Strengthening by Plastic Corrugated Reinforcements: An Efficient Way for Strain-Hardening Improvement by Architecture

O. Bouaziz, S. Allain , D. Barcelo, R. Niang
ArcelorMittal Research, Voie Romaine-BP30320, 57283 Maizières-lès-Metz Cedex, France

ABSTRACT

The particular plastic behaviour of corrugated metallic strips under tension has been investigated and it has been exploited to propose a new design of structural material using strengthening by plastic corrugated reinforcement. It is reported that the proposed approach is suitable to strongly improve the strain-hardening by this specific architecture.

INTRODUCTION

In material science the improvement of the strain-hardening is often a crucial challenge for the development of alloys or composites suitable to be formed without localization of the plastic strain. Unfortunately strain-hardening tends to decrease with an increasing strength. This problem has been widely investigated and metallurgical solutions such metal matrix composites [1,2], multiphase alloys (as steels [3] or Ti-based [4]) and using also dynamic hardening mechanism as Tranformation Induced Plasticity [5] or Twinning Induced Plasticity [6] have been proposed. As mentioned by pioneering work of Ashby [7] and as illustrated in a recent review [8] the metallurgical limitations to combine contradictory properties could be overcome changing the architecture of the materials at a scale between microstructure and samples or part. In this paper a method is proposed to combine more efficiently strength and strain-hardening. The approach is based on the interest of the use of plastic corrugated reinforcement embedded in a matrix has been investigated. In a first part the particular plastic behaviour of a corrugated metal under tension is characterized. In a second part the improvement of the matrix strengthened by corrugated reinforcements is reported.

 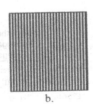 

a. b. c.

Fig1. Usual morphology of the reinforcement by particle (a.) and by fiber (b.) . Comparison with the approach based on corrugated reinforcement (c.)

EXPERIMENTAL PROCEDURE

Strips of Interstitial Free steel have been used. The thickness is 1 mm. Tensile specimen have been first machined to 20mm width and 80mm gauge length. In a second step different corrugations have been obtained by deformation at room temperature using uniaxial compression of the tensile specimen between two anvils where metallic cylinders have been added as shown in Fig.2a. This device is suitable to produce varying corrugation geometries changing the diameter of the cylinders and/or the spacing between cylinders. One of the obtained corrugated patterns is shown in Fig2.b.

The tensile behaviour have been characterized for the flat sample (i.e. without any corrugations) and for two samples with 2 and 4 corrugations with the geometry shown in Fig.2b. The experimental evolution of the engineering stress (force divided by the initial area) as a function of the elongation is drawn in Fig3. Initially the elongation of the corrugated samples is provided by un-bending of the corrugations. In the second step the behaviour of the flat product is achieved after corrugations have vanished. A change of curvature of the force-elongation curves is observed at the end of the first step for a critical value of the strain.

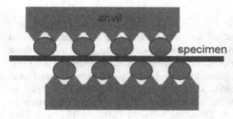
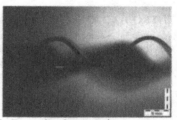

a. Device to make corrugated samples   b. Example of corrugation

Fig2. Fabrication of corrugated tensile samples

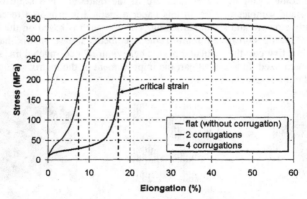

Fig3. Experimental evolution of the stress as a function of elongation for flat and corrugated strips

Using the simplified geometry (Fig.4a) it is easy to calculate order of magnitude values of the uniaxial strain provided for one corrugation of a wavelength p and an amplitude h after a complete unbending process :

(1)

$$\varepsilon_c = 2\sqrt{\left(\frac{h}{p}\right)^2 + \frac{1}{4}} - 1$$

Because the sample specimen of a gauge length L contains N corrugations in our case, the macroscopic average uniaxial elongation is :

(2)

$$\varepsilon_c^* = \varepsilon_c.N.\frac{p}{L}$$

For the selected corrugation geometry (Fig.2.b.) the ratio between the amplitude and the wavelength has been measured to be 0.28, the macroscopic uniaxial elongation induced by un-bending is drawn in Fig.4.b. as a function of the number of corrugations and compared with measurement from Fig3.

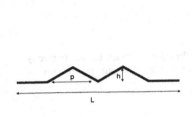

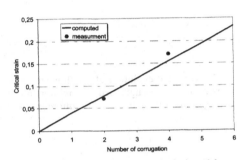

a. Schematized geometry of the corrugated tensile samples (p : the wavelength, h: the amplitude, L : the gauge length)

b. Computed and measured strain induced by un-bending of corrugations as a function of the number of corrugations.

Fig4. Determination of the uniaxial strain induced by un-bending process

SIMULATIONS

In order to check the possible exploitation of the specific behaviour of corrugated sheets in tension for an improvement of strain-hardening by corrugated reinforcement, numerical simulations by finite element method (Abaqus code) have been performed. Tensile tests along the Y-axis have been simulated on a unit cell with the geometry shown in Fig5. The simulations have been carried out in 2D i.e. the dimension along the Z-axis is assumed infinite and the volume fraction of reinforcement is 14%. The behaviours of the matrix and of the reinforcement are perfectly plastic with a respective yield stress of 250 and 600MPa. A Young modulus of 210GPa and a Poisson ratio of 0.3 are used for the elastic law. Fig6 highlights the interest the corrugated reinforcements in order to master the strain-hardening. If the common straight reinforcement provides a strain-hardening vanishing rapidly, the corrugated cases show that it is clearly possible to have a increase in the strain-hardening as a function of the plastic strain in a range of deformation depending of the corrugation geometries.

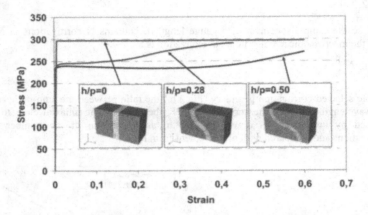

Fig5. Computed evolution of the flow stress as a function of strain for the perfectly plastic matrix strengthened by reinforcement with the different geometries of corrugation (volume fraction of reinforcement of 14%)

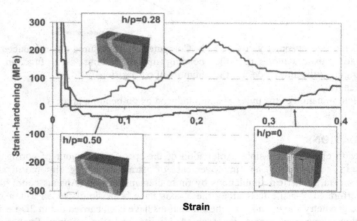

Fig6. Strain hardening as a function of strain for the perfectly plastic matrix strengthened by reinforcement with the different geometries of corrugation

CONCLUSION

Usually strain-hardening is a decreasing function of the plastic strain or of the flow stress. By using corrugated plastic reinforcements it is shown that an increase in strain-hardening with deformation could be achieved illustrating another interest of the architectured materials. The further developments of this approach are :
- a better combination between corrugated reinforcement and matrix,
- the process to produce such materials,
- an extension to have the same beneficial effect whatever the strain path i.e. the 2D extension of the concept.

REFERENCES

[1] J.D. Atkinson, L.M. Brown, M.M. Stobbs: Mat. Sci. Eng. vol.5 (1969) 193.

[2] D.J. Lahaie, J.D.Embury, M.F.Ashby: Scripta Met. vol.32 (1995) 133.

[3] M. Delincé, Y. Bréchet, J.D. Embury, M.G.D. Geers: P.J. Jacques, T. Pardoen, Acta Mat. vol.55 (2007) 2337.

[4] A. Sreeramamurthy, H. Margolin: Met. Trans. A vol.13 (1982) 595.

[5] P. Jacques, Q. Furnémont, A. Mertens, F. Delannay: Phil. Mag. A vol.81 (2001) 1789.

[6] O. Bouaziz, S. Allain, C. Scott: Scripta Mat. vol.58 (2008) 246.

[7] M.F. Ashby, Y. Brechet: Acta Mat. vol.51 (2003), 5801.

[8] O. Bouaziz, Y. Bréchet, J.D. Embury: Adv. Eng. Mat. vol.23 (2008) 1121.

Mater. Res. Soc. Symp. Proc. Vol. 1188 © 2009 Materials Research Society 1188-LL05-06

Topological Interlocking in Design of Structures and Materials

Yuri Estrin[1,2], Arcady Dyskin[3], Elena Pasternak[4] and Stephan Schaare[5]

[1]ARC Centre of Excellence for Design in Light Metals, Department of Materials Engineering, Monash University, Clayton, VIC 3800, Australia

[2]CSIRO Division of Materials Science and Engineering, Clayton, Vic. 3168, Australia

[3]School of Civil and Resource Engineering, The University of Western Australia, 35 Stirling Highway, Crawley, WA 6009, Australia

[4]School of Mechanical Engineering, The University of Western Australia, 35 Stirling Highway, Crawley WA 6009, Australia

[5]Rheinmetall Landsysteme GmbH, Kassel, Germany

ABSTRACT

Since its introduction in 2001 [1], the concept of topological interlocking has advanced to reasonable maturity, and various research groups have now adopted it as a promising avenue for developing novel structures and materials with unusual mechanical properties. In this paper, we review the known geometries of building blocks and their arrangements that permit topological interlocking. Their properties relating to stiffness, fracture resistance and damping are discussed on the basis of experimental evidence and modeling results. An outlook to prospective engineering applications is also given.

INTRODUCTION

In a quest for hybrid materials and structures providing multifunctionality along with favorable mechanical properties, materials researchers are increasingly turning their attention to geometry inspired designs [2]. One of the promising design principles is that of topological interlocking [1,3,4]. The rationale behind this principle is as follows. For brittle materials, there are obvious benefits of using fragmented, rather than monolithic structures. Indeed, due to the size effect, as represented, e.g. by the Weibull statistics [5], the failure probability of a structure is much higher than that of its constituent elements. Should it become possible to break a massive body down to small building blocks and then re-construct it from the fragments, a structure with a much higher resistance to failure would be obtained – provided, of course, that the building blocks can be held together in an efficient way to provide structural integrity of the re-constructed body. In our earlier work [1,3-7] we claimed that this can be rendered possible through the use of specific geometrical shapes and arrangements of the blocks, which ensure their geometric, or topological interlocking. According to this concept, an individual element is held in place by its neighbors kinematically, rather than through connectors or binder. Imagine a wall from which a brick cannot be removed not because it is attached to its neighbors by mortar, but just because its shape and the shapes of the neighbors precludes that. This is the principle that has been implicitly used in construction of dry stone walls in medieval Europe [8], fortified structures of Japanese castles [9], Incan masonry [10], etc. So the principle is not new, and its viability has been tested over centuries. What is new in our design is that the building blocks are *identical in shape and size*, which are specifically engineered to meet particular requirements. We have developed several types, or 'families', of the geometrical shapes that permit interlocking and have done mechanical tests demonstrating unusual mechanical properties of prototype structures based on topological interlocking of their elements. In what follows, we shall describe the possible interlocking geometries, present selected experimental results obtained with interlocked structures and give an outlook to possible technological applications of this design principle.

TYPES OF INTERLOCKING GEOMETRIES

Osteomorphic blocks

A simple geometry of a building block ensuring interlocking with its neighbors utilizes matching their concavo-convex surfaces, as shown in figure 1 where an exploded view of a group of 3 osteomorphic blocks is given. The term 'osteomorphic' refers to the bone-like shape of a block [5]. A planar assembly of such blocks is seen in figure 1 (right). When the blocks are brought into contact, the concave parts of each block match the convex parts of its six neighbors in such a way that its movements in any direction, particularly the one normal to the exposed flat surfaces of the blocks, are inhibited. The blocks are designed in such a way that a rectangular plate can be completed on its edges by halves of standard blocks cut along one of the two symmetry planes. These half-blocks are shown in figure 1 along with a full block. An analytical function describing the shape of the concavo-convex surfaces of osteomorphic blocks was given in Ref. 5. Modifications of the surface profile preserving the interlocking property are, of course, possible and, in fact, have been considered.

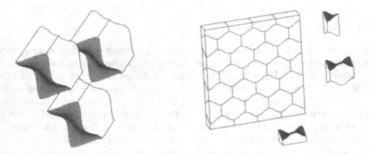

Figure 1. Geometry of an osteomorphic block and principle of assembly of osteomorphic blocks to a planar structure.

Obviously, the blocks at the edges of the plate do not have the right topological environment and need to be constrained – either externally, through a frame (figure 2) or, to avoid the use of a frame, through tensioned cables or wires (figure 3). Experimental studies on assemblies of osteomorphic blocks made from casting resin [5] and concrete [11] have demonstrated one of the principal benefits of the topological interlocking design: an increased resistance of structures engineered in this way to crack propagation under concentrated quasistatic loading. Whereas major macrocracks lethal to a monolithic plate were observed in such conditions, only a few osteomorphic elements adjacent to the indenter failed in a plate assembled from osteomorphic blocks [5]. The main reason for that is that a crack generated within an element gets blunted at an interface and cannot propagate through the segmented structure in a catastrophic way. In a situation where such an assembly is exposed to multiple local loading events (e.g. projectile impacts) at random, failure of a few individual elements can easily be tolerated by the structure – as a matter of fact, simulations have shown that over 24% of the elements would have to be destroyed before a percolation limit is reached and a backbone cluster of broken elements is formed [12]. This property makes assemblies of interlocked osteomorphic blocks particularly attractive for applications as protective layers or walls. It was suggested [6] that the use of interlocked non-planar tiles (e.g. modified osteomorphic blocks) in a spacecraft of the Columbia type would have provided a better resistance to failure.

Figure 2. Assembly of osteomorphic blocks with external constraint through a laterally loaded frame [5].

A further benefit of the topological interlocking design is the possibility to mix various materials within a hybrid structure – virtually in any combination and/or proportion – provided the elements made from different materials are of the same shape and size. The particular geometry of the osteomorphic block shown in figure 1, with its length being twice its width, makes it possible to construct corner structures, pillars, chimneys, etc., in which the elements are interlocked [5]. Furthermore, by placing a pair of twin blocks normally to the wall plane and building a next layer around them, as shown in figure 4, a structure consisting of two layers of interlocked elements, with an interlocked 'bridge' between them is created. In constructing such two-layer structures a desired number of bridging elements can be built in at will. Obviously, a multi-layer structure with an arbitrary large number of layers can be built in this way. This is a possibility we did not present in our original work, and the properties of such interlocked multi-layer structures will be studied in future research.

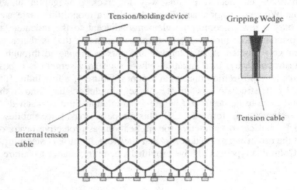

Figure 3. Assembly of osteomorphic blocks with external constraint via tensioned cables [4].

Figure 4. Out-of-plane pair of 'twin blocks' (seen in grey) creating a bridge between two layers of interlocked elements.

Convex polyhedra

A second 'family' of elements, which can be assembled in a structure where they are fully interlocked, derives from platonic bodies. It has been shown [13] that all five platonic shapes (tetrahedron, cube, octahedron, dodecahedron and icosahedron) can give rise to interlocked assemblies. Historically, the tetrahedron was the first platonic body we looked at in the context of topological interlocking [1]. Based on a consideration of the evolution of a cross-section through a layer-like array of interlocked tetrahedra (figure 5) as it moves away from the middle section through the layer, we developed a principle for constructing interlocking arrangements of the other platonic solids. In the case of tetrahedral elements, the middle section plane through the structure (figure 5) is fully tiled with squares. Tiling of the middle plane with regular hexagons permits construction of a layer of interlocked cubes, octahedra or dodecahedra [13]. In simple terms this principle of building up an interlocked structure from the middle plane up and down can be illustrated by figure 5, which shows the transformation of the tiling in the middle section of an arrangement of interlocked elements when the section is moved parallel to the middle one. The figure demonstrates that for the case when interlocking exists, the middle section of an element cannot fit into the 'window' formed by the sections of its neighbours in a plane parallel to the middle one. A criterion for interlocking of convex polyhedra was given in Ref. 5 by considering the polygons formed by the intersections of the extensions of an element faces constrained by the element's neighbours with a section plane parallel to the layer. It can be expressed as follows: An element is locked within the layer if, and only if, by continuously shifting the section plane in either direction with respect to the middle plane the corresponding polygon eventually degenerates to a segment or a point. A rigorous mathematical formulation of the principles guiding rational design of topologically interlockable elements was presented in Ref. 14. An alternative way of achieving interlocking with dodecahedra (as well as with icosahedra and truncated icosahedra – buckyballs) was also proposed [13].

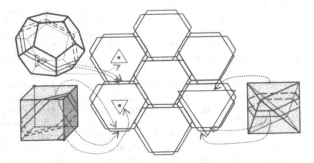

Figure 5. Transformation of the honeycomb structure in the middle section through an assembly of interlocked cubes, octahedral or dodecahedra (regular hexagons shown in magenta) with parallel translation of the section plane (distorted hexagons shown in blue). Note that these sections degenerate to a triangle (in the case of octahedra) or solid dots (in the case of cubes and dodecahedra) [13].

Figure 6. Interlocked octahedral blocks (left) and assembly of interlocked cubes under concentrated load (right).

Examples of assemblies of interlocked octahedra and cubes built up by transforming the middle section with upward or downward parallel displacement in the way described above are shown in figure 6. From the viewpoint of ease of manufacturing, assemblies of interlocked cube-shaped elements are of interest, although they do not show the tolerance to local failures, which structures of tetrahedral elements do [13]. The mechanical response of a layer consisting of topologically interlocked cubes was investigated experimentally and simulated numerically [15]. A very unusual feature, *viz.* a negative stiffness in the unloading part of the load *vs.* indenter displacement curve was found. This phenomenon was associated with the change of the type of contact between the cube-shaped blocks from the surface-to-surface to the apex-to-surface contact.

Interlocked arrangements of cube-shaped elements were used for various experiments aimed at testing such aspects of interlocked structures as the mentioned possibility of producing a desired materials mix and studying its mechanical response or investigating the energy absorption capacity of a layer of interlocked cubes. Here we present just three snippets from the results produced in this way. Figure 7 shows the dependence of the stiffness of a random mix of interlocked cubes-shaped PVC and steel elements (as illustrated by the left-hand side picture) on the fractions of the two species. It is obvious that a simple rule of mixtures does not apply to this behaviour.

A simplified model can be based on the assumption that the overall plate bending stiffness that controls the relation between the indentation force and the deflection is controlled by the effective modulus of the mixture of the PVC/steel cubes. The effective modulus of the mixture will be approximated by the readily available model of the effective modulus of interacting spherical inclusions embedded in a matrix [16]. The model is expressed by a set of differential equations

$$\begin{cases} \dfrac{d\kappa}{dc_1} = \dfrac{\kappa_1 - \kappa}{1 - c_1}\dfrac{\kappa + \kappa^*}{\kappa_1 + \kappa^*}, & \kappa^* = \dfrac{4}{3}\mu \\[2ex] \dfrac{d\mu}{dc_1} = \dfrac{\mu_1 - \mu}{1 - c_1}\dfrac{\mu + \mu^*}{\mu_1 + \mu^*}, & \mu^* = \dfrac{1}{6}\mu\dfrac{9\kappa + 8\mu}{\kappa + 2\mu} \\[2ex] \kappa\big|_{c_1 = 0} = \kappa_2 \\[1ex] \mu\big|_{c_1 = 0} = \mu_2 \end{cases}$$

where, κ and μ, are the effective bulk modulus and shear modulus and κ_1, μ_1, κ_2, μ_2, are the bulk moduli and shear moduli of the inclusions and the matrix, respectively; c_1 is the volumetric fraction of inclusions. These moduli can be expressed through Young's modulus E and Poisson's ratio v:

$$\kappa = \frac{E}{3(1 - 2v)}, \quad \mu = \frac{E}{2(1 + v)}$$

A solution of this set of equations is shown in figure 7 (right) by a solid line. It was obtained by assuming that the measured stiffness is proportional to the effective modulus of the assembly. Subsequently, by relating the stiffness to the modulus for a pure PVC assembly in the same way, the proportionality factor between the stiffness and the modulus can be eliminated. The ratio of the effective stiffness of the assembly to that of an assembly of PVC blocks can thus be represented by the ratio of the effective modulus of the assembly to that of the PVC one. Correspondingly, the relative modulus of steel (normalized with respect to that of PVC) was taken as the ratio of the stiffnesses of the full steel and full PVC assemblies. Poisson's ratio was taken as 0.3 for both steel and PVC cubes. (It should be noted that its influence is marginal.)

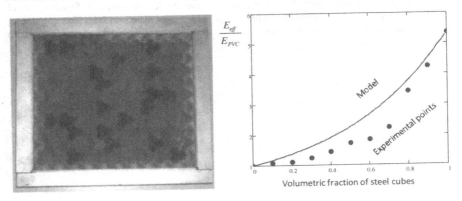

Figure 7. A random arrangement of interlocked steel (20%) and PVC (80%) cubes (left) and the dependence of the stiffness (defined as a secant modulus of the force-deflection curve) on the composition of the assembly (right).

123

It is seen that while the maximum error reaches 43% the model qualitatively captures the influence of mixing blocks made of materials whose moduli differ by almost an order of magnitude. Of course, the replacement of a 2D assembly of interlocking cubes with spheres in a 3D matrix can be one of the sources of the inaccuracy. Another source of error is obviously the relatively small number of cubes involved in the experiments, while the theory of effective characteristics assumes an unbounded matrix with infinite number of inclusions.

A second result pertaining to an assembly of cubes, which is worth mentioning, is its damping behaviour [17]. A typical load-displacement curve for quasi-static concentrated loading for an assembly of 100 Al cubes under a lateral load of 1.5 kN on the constraining frame is shown in figure 8 (left). An obvious hysteresis and a negative slope of the curve in the unloading part are seen. A comparison with the simulated curve suggests that the coefficient of friction μ between the Al blocks was about 0.3 [17]. The calculated dependence of the damping capacity ($\Psi=\Delta W/W$), defined as the relative dissipated part of the mechanical work, i.e. as the ratio of the area of the hysteresis loop, ΔW, to the area W below the loading curve, is also shown in figure 8. The diagram was obtained using the numerical model described in [15], which was verified experimentally. The very large values of dissipated energy are remarkable. The very large values of dissipated energy are remarkable. They suggest that interlocked structures may be efficiently used for energy absorption. Surprisingly, the extrapolation of the curve to zero coefficient of friction corresponds to a nonzero damping. A very good correspondence between the numerical simulations and the measurements of damping in the range where the coefficient of friction was known [17] does not rule out that the limit case of zero μ may not be represented by the simulation adequately because of fictitious 'energy dissipation' due to numerical discretization. However, the result may be genuine, as energy losses may be associated with channelled energy transport along the close-packed directions in the cube structure to the constraining frame where it gets dissipated. It should be mentioned in this connection that Autruffe et al. [18] have looked at the effect of friction on energy dissipation by using assemblies of osteomorphic blocks made from ice and changing μ simply through temperature variation. Ψ was shown to drop with decreasing μ, which is comparable with the behaviour of the low μ branch on figure 8, but, of course, in their real experiments the limit case of zero μ was never reached. The question of whether other energy dissipation channels than friction between the blocks exist thus remains unresolved at present.

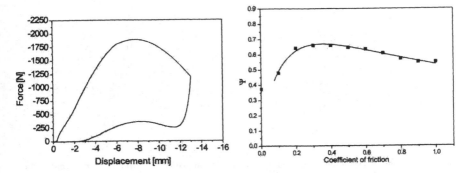

Figure 8. Hysteresis in the loading-unloading behaviour of an assembly of 100 interlocked Al cubes (left) and the calculated dependence of damping on the coefficient of friction (right) [17].

Finally, the maximum load bearing capacity of an assembly of interlocked cubes [15,17] (and, for that matter, any other convex polyhedra) and its stiffness increase with the magnitude of the lateral load exerted on the assembly through the outer frame or other tensioning devices. This offers an interesting possibility of controlling the bending rigidity of a layer of topologically interlocked elements of the structure, e.g. by actuating the lateral load.

Tubular elements

A third family of interlockable elements we wish to consider here are the tubular ones. As was shown in Ref. 7, a series of derivates of assemblies of interlocked tetrahedra can be developed by transforming the basic tiling pattern of the elements in the middle plane. Thus changing the squares (tiling pattern of tetrahedra) to octagons leads to an element shown in figure 9. Such elements are interlockable. Continuing the process of doubling the number of vertices of the polygons in the middle plane we get, as a limit case, a distorted cylinder with a circular cross-section in the middle plane, figure 10. In a cross-section away from the middle, an elliptical shape is seen, the aspect ratio growing with the distance from the middle plane. The ellipse at the top of this body can be seen as a result of rotation of that at the bottom by 90°. The cylinders can be made hollow giving rise to an assembly of interlocked twisted tubes, figure 11. The load bearing capacity of such a structure, particularly in the limit case of circular cross-sections, is, of course, very low, but it may be sufficient for applications, which do not require high loads. We proposed to manufacture such assemblies using selective laser sintering [19], rather than assembly of individual elements. It is also possible to use tubular 'distorted cylinder' elements in which the aspect ratio of the cross-sections varies periodically along the axis, so that the basic elements seen in figures 10 and 11 are continuously repeated in the axial direction. Interlocked bundles of such twisted tubes, once constrained laterally, may have a range of applications, see below.

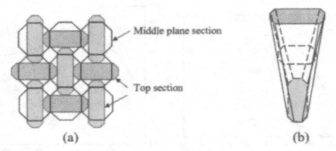

Figure 9. Interlocked elements based on truncated octagon prisms [7]: (a) the arrangement in the central section and the top section; (b) sketch of the prism.

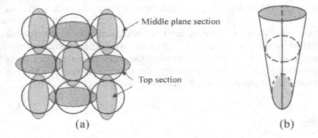

Figure 10. Interlocked elements based on twisted cylinders [7]: (a) the arrangement in the central section and the top section; (b) sketch of the cylinder.

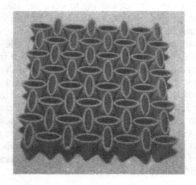

Figure 11. Interlocked tubular elements produced by selective laser sintering [19].

DISCUSSION

The above examples of geometry inspired design based on topological interlocking of identical elements represent one of the possible strategies for developing architecturally engineered materials and structures [20]. Interestingly, this principle, which appears attractive for manufacturing of man-made structures, is rarely realized in living matter. One exception is the recently observed structure of the soft suture between bone segments in the shell of a turtle, where geometrical interlocking may be actuated to increase stiffness [21]. The possibility of controlling the load bearing capacity or bending rigidity of artificial structures consisting of interlocked elements of the kind presented is certainly a very attractive one.

We see the main applications of the interlocking-based structures in such areas as civil engineering, where the high resistance to crack propagation, large energy absorption capacity and remarkable tolerance to local failures can be used both for construction and for protection of existing structures and personnel. Construction in earth-quakes affected regions is a further application of interest, as is the vast area of noise protection. We have also identified a number of applications of interlocking-based design in extraterrestrial construction [4]. Such aspects as the possibility to combine various materials (and hence functionalities) within a single topologically interlocked structure as well as ease of assembly and dismantling/recycling provide additional benefits.

The scale independence of topological interlocking design makes it possible in principle to downscale the structures based on this concept for use as reinforcement in composites. Should a manufacturing technique become available for producing and assembling micro- or nanoscale interlocking elements, a broad range of applications in protective surface layers would emerge. We believe that a most viable way of producing such structures would be through layer-by-layer build-up, which may utilise lithography or other programmable layer-by-layer deposition techniques.

Tubular elements described above also deserve attention. Due to the bizarre, tortuous shape of the inner channel they have, their use for sound attenuation or breaking of gas or fluid flow is envisaged. The tubular interlockable elements can be manufactured to possess a very high surface to weight ratio, particularly if a degree of porosity is deliberately factored in, e.g. in a selective laser sintering process [19]. These features can be used for catalytic reactions, in drug-eluting structures, etc.

CONCLUSION

In this overview we have presented topological interlocking as a design principle and gave a few examples of its possible use in novel structures and materials. A creative engineer will certainly find further areas of applications of this geometry-inspired design principle. The main issue with implementing the proposed geometries is the manufacturability of the individual building blocks and/or entire assemblies of interlocked elements, and we trust that with the advent of viable engineering solutions for the manufacturing of segmented structures and materials based on topological interlocking the ideas outlined here will start finding their way to industrial applications.

ACKNOWLEDGMENTS

The authors are grateful to Profs. Y. Brechet and M.F. Ashby for their encouragement and useful discussions. Thanks are due to David Yong for providing information on the dry stone cases. YE acknowledges partial support from DFG through grant Es 74/10. Support from the Australian Research Council through the Discovery Grants DP0559737 (AVD) and DP0988449 (AVD, EP) is also acknowledged.

REFERENCES

1. A. V. Dyskin, Y. Estrin, A. Kanel-Belov and E. Pasternak, *Scripta mater.* **44**, 2689-2694 (2001).
2. M. F. Ashby, Y. J. M. Bréchet, *Acta Mater.* **51**, 5801 (2003).
3. A. V. Dyskin, Y. Estrin, A. J. Kanel-Belov and E. Pasternak, *Adv. Eng. Mater.* **3**, 885-888 (2001).
4. A. V. Dyskin, Y. Estrin, E. Pasternak, H. C. Khor and A. J. Kanel-Belov, *Acta Astronautica* **57**, 10-21 (2005).
5. A. V. Dyskin, Y. Estrin, E. Pasternak, H. C. Khor and A. J. Kanel-Belov, Adv. Eng. Mater. 5, No. 3, C307 (2003).
6. Y. Estrin, A. V. Dyskin, E. Pasternak, H. C. Khor, A. J. Kanel-Belov, *Phil. Mag. Letters* **83**, 351 (2003)
7. A. V. Dyskin, Y. Estrin, A. J. Kanel-Belov, E. Pasternak, *Comp. Sci. Techn.* **63**, 483–491 (2003).
8. A. Radford, *A Guide to Dry Stone Walling*, (Crowood Press, Ramsbury Wiltshire, UK, 2002).
9. R. Hoshino, In: *Civil Engineering in the unearthing and preserving of historic ruins*, (Japan Society of Civil Engineers, 2000), pp. 18-21.
10. J. Hyslop, *Inka settlement planning*, (University of Texas Press, Austin, USA, 1990).
11. H. C. Khor, A. V. Dyskin, Y. Estrin and E. Pasternak, In: *Structural Integrity and Fracture SIF 2004*, edited by A. Atrens, J. N. Boland, R. Clegg and J. R. Griffiths, (Brisbane, 2004) pp. 189-194.
12. A. Molotnikov, Y. Estrin, A. V. Dyskin, E. Pasternak, A. Kanel-Belov, *Eng. Fract. Mech.* **74**, 1222-1232 (2007).
13. A. V. Dyskin, Y. Estrin, A. J. Kanel-Belov, E. Pasternak, *Phil. Mag. Letters* **83**, 197-203 (2003).
14. A. J. Kanel-Belov, A. V. Dyskin, Y. Estrin, E. Pasternak, I. A. Ivanov-Pogodaev, *Moscow Mathematical Journal* (2009, in press).
15. S. Schaare, A. V. Dyskin, Y. Estrin, S. Arndt, E. Pasternak, A. Kanel-Belov, *Intl. J. Eng. Sci.* **46**, 1228-1238 (2008).
16. R. McLaughlin, *Intern. J Engineering Science*, **15**, 237 -244 (1977).
17. S. Schaare, *Charakterisierung und Simulation topologisch verzahnter Strukturen*, PhD Thesis, (TU Clausthal, 2008).
18. A. Autruffe, F. Pelloux, C. Brugger, P. Duval, Y. Bréchet, M. Fivel, *Adv. Eng. Mater.* **9**, 664-666 (2007).
19. Y. Estrin, N. Müller, D. Trenke, A. Dyskin, E. Pasternak, *Structure Composed of Elements and Method of its Production*, US Patent #6884486 (26.4.2005).

20. O. Bouaziz, Y. Bréchet, J.D. Embury, *Adv. Eng. Mater.* **10**, 24-36 (2008).
21. S. Krauss, E. Monsonego-Ornan, E. Zelzer, P. Fratzl, and R. Shahar, *Adv. Mater.* **21**, 407-412 (2009).

Mater. Res. Soc. Symp. Proc. Vol. 1188 © 2009 Materials Research Society   1188-LL05-02

Design of Architectured Sandwich Core Materials Using Topological Optimization Methods

Laurent Laszczyk[1], Rémy Dendievel[1], Olivier Bouaziz[2], Yves Bréchet[1] and Guillaume Parry[1]
[1]SIMAP, Grenoble INP, CNRS, UJF, 101 rue de la physique BP46, 38402 St-Martin-d'Hères Cedex, France
[2]ArcelorMittal Research. Voie Romaine BP30320, 57283 Maizières-lès-Metz Cedex, France

ABSTRACT

Sandwich structures are especially interesting when multiple functionalities (such as stiffness and thermal insulation) are required. Properties of these structures are strongly dependent on the general geometry of the sandwich, but also on the detailed patterns of matter partitioning within the core. Therefore it seems possible to tailor the core pattern in order to obtain the desired properties. But multi-functional specifications and the infinite number of possible shapes, leads to non-trivial selection and/or optimization problems.

In this context of "material by design", we propose a numerical approach, based on structural optimization techniques, to find the core pattern that leads to the best performances for a given set of conflicting specifications. The distribution of matter is defined thanks to a level-set function, and the convergence toward the optimized pattern is performed through the evolution of this function on a fixed grid. It is shown that the solutions of optimization are strongly dependent on the formulation of the problem, which have to be chosen with respect to the physics.

A first application of this approach is presented for the design of sandwich core materials, in order to obtain the best compromise between flexural stiffness and relative density. The influence of both the initialization (starting geometry) and the formulation of the optimization problem are detailed.

INTRODUCTION

Specifications in automotive industries are more and more complex in the way that multi-functional performances (such as stiffness, thermal and acoustic insulation) are required while keeping the weight as low as possible. Composite and architectured materials often give efficient solutions. Recently, numerous techniques have been developed for manufacturing lightweight structures like graded metal foams [1], hollow spheres and trusses [2,3], corrugated structures, honeycombs and other cellular solids [4]. In a "material by design" approach [5], finding the cellular architecture that combines in the most efficient way a given set of conflicting properties is the key issue. In this regard, shape optimization techniques can enlarge the degrees of freedom on the geometry.

In this paper, topological optimization is applied to periodically architectured flexural panels. Before any comparison or optimization, it is necessary to define a way to measure the performances of those panels. In a first part, a determination method of the effective properties is presented on three test architectures. Then, a validation of these effective properties is numerically done on a four-point bending test. The bending stiffness of the homogeneous effective panels is then compared to FEM simulations of the whole architectured panels. Then in a second part, topological optimization by level-set method [6] is shortly described, and different optimized geometries depending on the formulation of the optimization problem are presented.

EFFECTIVE PROPERTIES

A panel under bending is mainly subjected to flexural and shear loadings. The mechanical behavior of the architectured panels can then be described using an effective flexion and shear modulus. The numerical determination of these effective moduli is performed on three test architectures of same density (squares, triangles and porous pattern), as shown on Figure 1. Then, the four-point bending stiffness of the homogeneous effective panels is compared with the one numerically computed on the whole architectured panels.

Figure 1 Three test architectures: squares, triangles and porous patterns.

Numerical determination of effective modulus on the periodic unit cell

The panels under consideration are invariant in one direction and periodic in the other. Then, the effective moduli can be computed by a 2D plain strain finite element analysis on the unit cell [8] using the energy approach described below. The unit cell is a parallelepiped of length l, height h and width w.

For the flexion modulus calculation, a mean curvature χ is imposed, such as the strain appears as the sum of a curvature and a periodic term, as $\varepsilon = \chi y\, e_{xx} + \varepsilon^{per}$. Periodic conditions are applied on the right and left boundaries, while the upper and lower faces are kept unloaded. The effective flexion modulus is defined as the Young's modulus of an homogeneous material that leads to the same strain energy W_f^*:

$$\tilde{E} = \frac{24}{lh^3} \frac{W_f^*}{\chi^2}$$

For the shear modulus calculation, periodic conditions are also applied on the right and left boundaries, and displacements are imposed on the upper and lower faces. The upper, resp. lower, face is subjected to a displacement $u_x = \delta$, resp. $u_x = 0$. As previously, the expression of the strain energy of an homogeneous material gives the following effective shear modulus:

$$\tilde{G} = \frac{2h}{l} \frac{W_s^*}{\delta^2}$$

The effective moduli computed for the three test patterns are presented in Table 1. The moduli of the constitutive material (steel) are given for comparison. For an almost identical relative density of $\rho_r = 0,4$, the square and triangle patterns show a similar behavior, since only a small difference of shear modulus is observed. However, the porous pattern is really weaker with one order of magnitude lower than the constitutive material.

	$\tilde{E}$ [GPa]	$\tilde{G}$ [GPa]	Relative density ρ_r
Unstructured core	200	75,2	1
Square pattern	142	2,7	0,441
Triangle pattern	127	9,4	0,427
Porous pattern	47	4,4	0,435

Table 1 Effective moduli computed on the unit cell for the three test architectures, compared with the one of the constitutive material (steel).

Numerical validation with a four-point bending test

To validate the previous calculations on the unit cells, the comparison with the mechanical behavior of the whole architectured panels is necessary. The four-point bending test, which is often used to characterize flexural panels, was simulated for the previous test architectures. Figure 2 shows the result of simulation for the triangle patterned panel.

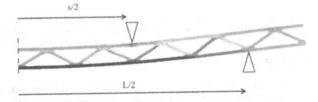

Figure 2 Simulation of the four-point bending test applied to the triangle patterned panel. Colors represent tensile stress σ_{xx}.

Then, comparison is done between the bending test simulations and the effective homogeneous panels. The values reported on Figure 3 are the four-point bending stiffness normalized by the width w, on one hand computed by simulation of the whole architectured panels, and on the other hand calculated using the analytical expression for an homogeneous material [9]:

$$R_f = \frac{F}{wd} = \left(\frac{(L-s)^2(L+2s)}{4h^3\tilde{E}} + \frac{L-s}{4h\tilde{G}} \right)^{-1}$$

where F is the applied load, d the displacement of the inner upper indenters, and $\tilde{E}$, $\tilde{G}$ the effective moduli previously computed on the unit cells.

The effective properties seem to give a relevant measure of the global flexural performance of the different panels. However, whereas a really good estimation is done for the porous pattern, the ones of the square and triangle patterns are more rough. This relative mismatch could be induced by the non separation of the scales necessary for homogenization, and also by the high ratio observed between flexion and shear modulus. The non separation of scales means that the size of the heterogeneities is close to the one of the variations of the strain or stress. A consequence of this is the flexion of the upper face just below the upper inner indenters, as shown Figure 2.

Finally, the previously presented method of calculation of the effective moduli on the unit cell gives relevant informations on the flexural performance of the architectured panels, even if local behaviors are omitted. Thus, it can be used as an objective function for an optimization procedure.

133

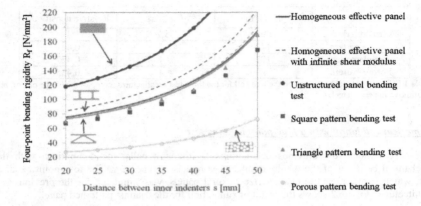

Figure 3 Four-point bending stiffness calculated for the homogeneous effective panel (with and without infinite shear modulus) in comparison with the one computed by bending test simulation. Unstructured core and the three test patterns are presented.

SHAPE OPTIMIZATION

Topological optimization techniques are based on numerical analysis and are usually applied to structures under mechanical or thermal loads. The computation of the physics PDE equations (usually linear mechanical or thermal equations) is done by finite element or finite difference analysis. Among numerous possible topology optimization techniques [8], a level-set method coupled with a shape gradient algorithm [6] was used. The technical aspect of the optimization algorithm are not described here, discussion is focused on how they can be applied to a multi-objective optimization of the previously defined architectured panels.

Level-set method

The optimization problem will be expressed as the maximization of an objective function J (e.g. the effective properties) while keeping satisfied some constraint inequalities (e.g. the relative density lower or equal to a given value). A shape optimization algorithm is a numerical iterative method that transforms the geometry at each iteration such as the objective function increases with respect to the constraints.

The level-set method means that the geometry is defined by a so-called level-set function, which is negative where there is matter, equal to zero on the boundary, and positive otherwise. At each step, the optimization procedure consists in calculating the objective function and the constraints (by finite element method), then in transforming the level-set function solving a transport partial differential equation (by finite difference method). The velocity for the transport equation is taken equal to the opposite of the shape gradient, so that the modified geometry increases the objective function.

Results

Two different formulations of the optimization problem are proposed: the first with an objective function as a linear combination of the effective moduli, the second with an objective function as the four-point bending stiffness. For each formulation, the relative density is kept constant ($\rho_r = 0,4$).

The first and more intuitive formulation, inspired from multi-functional material selection, is to maximize a linear combination of the effective moduli. The objective function is written as:

$$J_1 = \alpha \tilde{E} + (1 - \alpha)\tilde{G}$$

Figures 4 and 5 show the topological evolution of the porous pattern during the optimization respectively for flexion modulus and shear modulus (i.e. $\alpha = 1$ and $\alpha = 0$). It has been found that for intermediate values of the exchange constant α, one modulus is systematically neglected versus the other one. The critical value of the exchange coefficient depends on the starting geometry. Finally, with such an approach it is impossible to well weight the contribution of flexion and shear. One of them becomes systematically predominant.

Figure 4 Topological evolution of the porous pattern that maximize J_1 with $\alpha = 1$. Iterations 0, 5, 10, 15 and 20.

Figure 5 Topological evolution of the porous pattern that maximize J_1 with $\alpha = 0$. Iterations 0, 10, 20, 30 and 39.

In regard of the first part, the optimization problem can be formulated in terms of four-point bending stiffness, i.e. $J_2 = R_f$. In such a way, the balance between flexion and shear is not fixed anymore by an arbitrary exchange coefficient, but imposed by the physics. Figure 6 shows the topological evolution of the porous pattern with this objective function with the following dimensions: $s = 50$ mm and $L = 80$ mm. A mixture of the previous optimized geometries is obtained.

Figure 6 Topological evolution of the porous pattern that maximize J_2 with $s = 50$ mm and $L = 80$ mm. Iterations 0, 10, 20, 30 and 39.

The optimized architecture gives an compromise between the effective flexion and shear moduli, as plotted Figure 7. The final values of these moduli leads to a four-point bending stiffness (for the given dimensions) around 200 N/mm^2, which is better than the square and triangle patterns but not more than the error between the homogenized panels and the bending test simulations. To conclude, the optimization results show that the square and triangle pattern are close to the optimal architecture, and that sandwich with a 45° truss core seems to be the most efficient architecture for such a bending load.

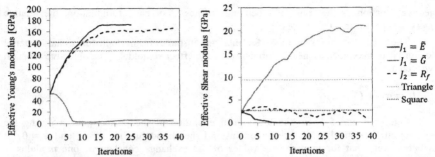

Figure 7 Evolution of the effective moduli during optimization for three different objective functions, compared to the one of the triangle and square patterns.

CONCLUSIONS

Flexural lightweight panels present numerous possible architectures that used to be compared and parametrically optimized. In order to generalize the optimization to all topologies, shape optimization was applied to 2D-periodic architectured panels under a bending load. Before optimizing, effective properties and the way to compute them on a unit cell had been presented. These effective properties have been calculated on three test patterns and compared to simulations of the whole architectured panels subjected to a four-point bending test. Results show that effective properties measure accurately the overall bending stiffness, even if local behaviors are omitted. Then, shape optimization have been done for two different objective function. It was shown that the four-point bending stiffness enable to physically weight the contributions of the flexion and the shear stiffness. Finally, the obtained optimized architecture and its flexural performances allow to conclude that the initial square and triangle patterns are close to the optimal solution, which seems to be a sandwich with 45° truss core. In ongoing work, multi-functional properties such as thermal insulation and stiffness are considered.

REFERENCES

1. A. Pollien, Y. Conde, L. Pambaguian and A. Mortensen. Graded open-cell aluminium foam core sandwich beams. *Materials Science and Engineering A*, **404**, 9-18, (2005).
2. P. Lhuissier, A. Fallet, L. Salvo and Y. Brechet. Quasistatic mechanical behaviour of stainless steel hollow sphere foam: Macroscopic properties and damage mechanisms followed by X-ray tomography. *Materials Letters*, **63**, 1113-1116, (2009).
3. D. T. Queheillalt and H. N. G. Wadley. Cellular metal lattices with hollow trusses. *Acta Materialia*, **53**, 303-313, (2005).
4. H. N. G. Wadley, N. A. Fleck and A. G. Evans. Fabrication and structural performance of periodic cellular metal sandwich structures. *Comp. Sc. and Tech.*, **63**, 2331-2343, (2003).
5. M. F. Ashby, and Y. J. M. Brechet. Designing hybrid materials. *Acta Materialia*, **51**, 5801-5821, (2003).
6. G. Allaire, F. Jouve and A. M. Toader. Structural optimization using sensitivity analysis and a level-set method. *Journal of Computational Physics*, **194**, 363-393, (2004).
7. N. Buannic, P. Cartraud and T. Quesnel. Homogenization of corrugated core sandwich panels. *Composite structures*, **59**, 299-312, (2003).
8. H. A. Eschenauer, and N. Olhoff. Topology optimization of continuum structures: a review. *Applied Mechanics Reviews*, **54**, 331-390, (2001).

Mater. Res. Soc. Symp. Proc. Vol. 1188 © 2009 Materials Research Society

Metallic Sandwich Structures With Hollow Spheres Foam Core

Pierre Lhuissier, Alexandre Fallet, Luc Salvo, Yves Bréchet, Marc Fivel
SIMaP-GPM2 - Grenoble Institute of Technology-CNRS-UJF, 101 rue de la Physique , BP46,
38402 Saint-Martin d'heres, France

ABSTRACT

Sandwich structures and foamed materials are typical architectured materials. Their combination provides potentially very performant solutions combining stiffness, strength, energy absorption and acoustic damping. The present contribution deals with the integration of a special type of foams, namely hollow spheres stackings, into sandwich structures.

Stainless steel hollow spheres main advantage relies on their smooth stress-strain curves and their very good repeatability, compared to other closed cell metallic foams. Therefore these foams are interesting alone but also in sandwich design.

A parametric study of the macroscopic behaviour of random stainless steel hollow spheres packing in uniaxial compression was carried out. Scaling laws for the Young's modulus, and for yield strength were established, and they are used to calculate sandwich properties.

Then one of the studied metallic hollow spheres packing has been integrated in a sandwich structure with stainless steel faces. Four point bending tests have been performed on various sandwich structures with four core thicknesses and three face thicknesses up to large deflection. We obtained thus the stiffness, the critical load where first damage occurs, the maximum load as a function of the sandwich parameters (core and face thickness). We compared this to classical analytical models.

INTRODUCTION

The integration of metallic foams in sandwich structures is of major interest when one is looking for a combination of properties such as stiffness, lightness, energy absorbed during damaging, high temperature working conditions and acoustic damping. Prior to the investigations of these properties, the ability to predict the behaviour of such structures where shown under various loading conditions : 3 and 4 points bending with clamped or simply supported boundaries conditions on beams [1-3] and indentation on plates. Various authors proposed failure modes maps [4]. Nevertheless few experiments have been done with hollow spheres foams core sandwich structures. In order to predict the failure mode of such structures an accurate knowledge of the core material properties is necessary. Some models have been developed to predict the behaviour of regular hollow spheres packing [5-7], but except for some works on few spheres [8] or purely experimental investigation [9], it seems that there is up to now a lack of quantification of the overall evolution of the properties of random hollow spheres stacking with the density. Thus compressive tests have been performed on stainless steel hollow spheres foams of various densities. Then four points bending tests have been performed on sandwich beams with hollow spheres foam core and with variable face thicknesses. Comparison between analytical predictions and experiments are then carried out both for the damaging strength and for the identification of the failure mode.

MATERIALS AND EXPERIMENTAL METHODS

Experiments have been performed both on the core material, and on the sandwich structures. Core material consists in a stacking of sintered stainless steel hollow spheres (314 norm AISI). For a given outer radius of 1.3 mm, the shell thickness ranges from 50 to 100 μm and thus the density is between 400 to 800 kg/m³. Simple compressive tests were performed on cubic samples of 30mm side size, made out materials with 3 differents densities. A classical compressive machine ADAMEL DY 35 was used with a 20kN load gauge. Displacement was measured thanks to a +/-10mm RDP sensor. Samples were unloaded frequently during the compressive test in order to obtain the evolution of the unloading modulus with the macroscopic strain.

Sandwich structures consists in a hollow spheres foam core of density 400g/L brazed on faces made with the same constitutive material (314 norm AISI). Sandwich structures were loaded in 4-point bending. The outer span was of 210mm while the inner span was of 100mm. Beams of 300 mm length and 50 mm width, cut by wire electro erosion, were simply supported on cylindrical rollers of radius 19 mm. Tests were performed on ADAMEL DY 35 with a 20kN load gauge. A +/- 10mm RDP sensor was used to measure the displacement of the central point of the structure during the first 16mm of deflection. Load-displacement curves were corrected to take in account the stiffness of the machine and the bending system.

EXPERIMENTAL RESULTS AND DISCUSSION

Hollow spheres foams behaviour

Compressive tests on hollow spheres foams were found to show a good repeatability. Furthermore, the macroscopic behaviour is traduced by a smooth curve with progressive variations with two regimes : a linear elastic regime and a post yielding linear hardening stage.
 Initial density only plays a role on the strength and hardening levels, but not significantly on the strain scale. Figure 1 shows stress-strain curves for 3 densities and how theses curves are conveniently described with bilinear laws.
The compressive stress strain curves are characterized by three quantities. The stress at 0.2% of plastic strain (obtained thanks to the loading modulus), the stress at 50% of strain, the initial unloading modulus, the mean hardening slope within the plateau region. Another caracteristic stress was defined as the intercept on the vertical axis of the plateau regime: this stress σ^* can be defined from the stress at 50% and the slope H as hiven in equation (1)

$$\sigma^* = \sigma_{50\%} - 0.5\,H \qquad\qquad (1)$$

The quantities were measured for materials with three initial densities and following the spirit of Gibson-Ashby laws [4] for cellular solids, power law expressions as function of the initial density were derived. Figure 2.a shows the scaling law for stress at 0.2% of plastic strain and for the extrapolated elastic strain σ^*, while Figure 2.b is relative to the initial unloading modulus. Table I summarises the exponent of the power laws for the various parameters considered.

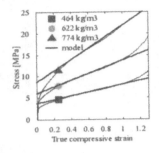

Figure 1. Strain-stress curves responses for compressive tests of various density foams with linear approximation model for the plateau.

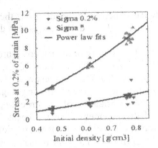 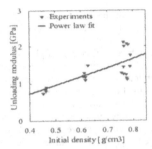

Figure 2. Experiments and power law fit for respectively **(a)** stresses at 0.2% of plastic strain ($\sigma0.2\%$) and elastic stresses extrapolated from the plateau (σ^*) and **(b)** for initial unloading moduli.

Property	Stress at 0.2% of plastic strain	Elastic stress extrapolated	Young modulus	Hardening modulus	Stress at 50% strain
Exponent n	1.55	1.80	1.21	2.12	1.92

Table I. Summary of the exponent of power law relating the characteristics of the stress-strain curve to the initial foam density.

Sandwich structures behaviour

From the four-points bending tests performed, three failure modes were observed : face yield, core shear and indentation. Figure 3 shows the evolution of the load and the thickness diminution of the sandwich versus displacement for 3 sandwich beams failing respectively by core shear, face yield and indentation. The thickness reduction of the sandwich is obtained thanks to the difference between displacements registered by the RDP sensor and the displacement imposed by the machine corrected by its stiffness. This parameter contains simultaneously the information of indentation and of curvature of the central part of the beam.

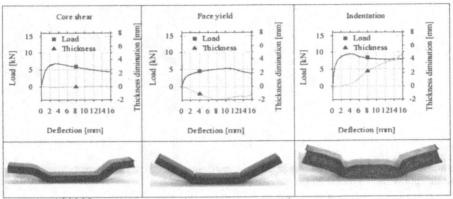

Figure 3. Comparison of the load responses of sandwich structures depending on their failure mode. The diminution of the total thickness of the structures with the deflection is also drawn.

Core shear induces formations of plastic hinges at the vicinity of the rollers, a mechanism which gives at the macroscopic scale a flat central part and no indentation. Core shear is therefore characterized by a constant sandwich thickness (Fig. 3a). The load response was shown to reach its maximum at low deflection and then the structure undergoes a softening stage.

Face yield is the signature of an important flexural deformation of the structure, thus the central deflection is more important than the imposed displacement. The thickness increases (Fig. 3b). The load-deflection curve shows a long hardening stage where faces are yielding. At maximal load the change of behaviour is more abrupt than for other failure modes. It is simply the results of instabilities that led to a localisation of the rotations of the sections of the beam. The shape of the structure at the end of the tests could be misinterpreted as an indentation failure, but the hardening stage clearly deny this hypothesis.

Indentation mode (Fig. 3c) is characterized by a limited displacement of the lower face while upper rollers penetrates the upper face of the structure. The thickness decreases. Maximal load can be reached at low deflection or at large deflection but whatever the case, the load remains relatively close to the maximal load during a large deflection stage.

Classical analytical models for theses failure modes [4] underestimate the properties of these sandwich structures. First faces hardening is not taken into account while the relatively important core thicknesses induced large strains in faces which will undergo workhardening. Furthermore, the core hardening is not included in the classical models, while it may also undergoes important strains. Last but not least, the assumption of rigid cross section within the theory of beams is challenged by the high deformability of the core. Because all these analytical models are based on the beam theory with elasto-perfectly plastic behaviour for the constitutive materials, their predictions are not surprisingly underevaluating the maximum strength. One has to keep in mind that theses failure models do not only look for the first damage within the structures, but they also look for the shape of damage. It induces that the failure is modelised thanks to the combination of various local damages of the structures such as plastic hinges, local plasticity of the core... As a consequence, if the level of stress when the failure occurs is underestimated, the type of the failure mode is quite well predicted, because the competition between the modes remains balanced. Figure 4 shows the locations of the tested structures in a failure map, where axis represents the ration c/L (core thickness over outer span) and t/c (face thickness over core thickness). For each structure the two curves as presented in figure 3 are shown close to the representative symbol of a given sandwich.

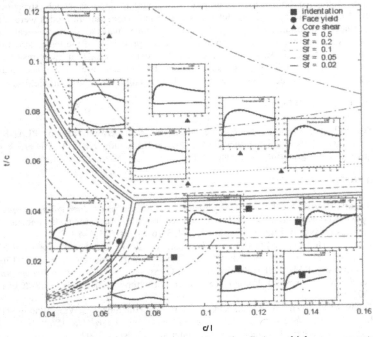

Figure 4. Failure map where axis represents the ratio c/L (core thickness over outer span) and t/c (face thickness over core thickness).

One can notice that when the safety factor Sf (that is an indication of the relative load distance to another damaging mode) is superior to 50%, the damaging mode is accurately predicted which may not be the case for lower values of Sf.. This is especially the case for core shear, for which the failure mode load is underestimated. If the stress at 0.2% of plastic strain had been used for the models, core shear would have been systematicaly predicted except for the 4 structures with 0.4 mm face thickness which would have been assumed to damage by indentation. By substituting in the models this yield stress by the extrapolated elastic stress σ*, the relative dominance of each failure modes is better accounted for. It remains however that core shear damaging load is sligthly underestimated.

CONCLUSIONS

From a parametric study among density of the stainless steel hollow spheres foams under compressive loading, scaling laws for representative mechanical properties have been extracted. Thus yield stress, Young modulus and hardening modulus can be extrapolated for any density within the range of the study (400g/L to 800 g/L). Since the compressive response was smooth and homothetic with density, an extrapolation of the behaviour with density is reasonable.

Four point bending tests on hollow sphere foam core sandwich structures were performed. Core thickness and face thickness were varied. Three failure modes were identified, and their relative importance was analysed using a failure map based on classical analytical models. It was shown that the models underestimate the properties of the tested structures, and that, if the yield stress is used to characterize the core materials, it may result in some discrepancies in the predictions of the failure modes. Failure maps are more in accordance with experiments if the foam is described using the crushing strength σ^* rather than yield strength..

ACKNOWLEDGMENTS

The hollow sphere materials and the sandwich plates were provided by PLANSEE®. It was initiated through the MAPO project funded by ONERA and CNRS. A PhD grant from the French government is gratefully acknowledged (PL).

REFERENCES

1. Tagarielli, V.L., Fleck, N.A. & Deshpande, V.S., *Collapse of clamped and simply supported composite sandwich beams in three-point bending*, Marine Composites, Composites Part B: Engineering, **2004**, Vol. 35(6-8), pp. 523-534
2. Chen, C., Harte, A. & Fleck, N.A., *The plastic collapse of sandwich beams with a metallic foam core*, Int. Journal of Mech. Sciences, **2001**, Vol. 43(6), pp. 1483-1506
3. McCormack, T.M., Miller, R., Kesler, O. & Gibson, L.J., *Failure of sandwich beams with metallic foam cores*, Int. Journal of Solids and Structures, **2001**, Vol. 38(28-29), pp. 4901-4920
4. Ashby, M.F., Evans, A., Fleck, N.A., Gibson, L.J., Hutchinson, J.W. & Wadley, H.N.G. *Metal foams: a design guide: Butterworth-Heinemann, Oxford, UK, ISBN 0-7506-7219-6, Published 2000, Hardback, 251 pp.*, Materials & Design, **2002**.
5. Gasser, S., Paun, F. & Brechet, Y., *Finite elements computation for the elastic properties of a regular stacking of hollow spheres*, Mat. Sc. and Eng. A, **2004**, Vol. 379(1-2), pp. 240-244
6. Sanders, W.S. & Gibson, L.J., *Mechanics of hollow sphere foams*, Mat. Sc. and Eng. A, **2003**, Vol. 347(1-2), pp. 70-85
7. Karagiozova, D., Yu, T. & Gao, Z., *Modelling of MHS cellular solid in large strains*, Int. Journal of Mech. Sciences, **2006**, Vol. 48(11), pp. 1273-1286
8. Lim, T.J., Smith, B. & McDowell, D.L., *Behavior of a random hollow sphere metal foam*, Acta Materialia, **2002**, Vol. 50(11), pp. 2867-2879
9. Friedl, O., Motz, C., Färber, J., Stoiber, M. & Pippan, R., *Tension and compression behaviour of stainless steel (316L) hollow sphere structures*, Cellular Metals and Polymers, **2004**.

Mater. Res. Soc. Symp. Proc. Vol. 1188 © 2009 Materials Research Society   1188-LL05-05

Numerical Simulations of Topologically Interlocked Materials Coupling DEM Methods and FEM Calculations: Comparison With Indentation Experiments

Charles Brugger[1,2], Marc C. Fivel[1] and Yves Brechet[1]
[1]SIMaP-GPM2, INP Grenoble/CNRS, 101 Rue de la Physique, BP46, 38402 St Martin d'Hères cedex, France
[2]Present address: Institute of mechanics, materials and civil engineering (iMMC),
Université catholique de Louvain, B-1348 Louvain-la-Neuve, Belgium

ABSTRACT

Planar assemblies of interlocked cubic blocs have been tested in indentation. Experiments are performed on blocs made of plaster. Influence of key parameters such as the surface roughness, the compression stress and the number of blocs are investigated. A numerical modeling is then proposed based on discrete element method. Each bloc is represented by its centre coordinates. Constitutive equations obtained by finite element simulations are introduced to model the contact between the blocs. The numerical tool is then applied to the case of indentation loading. It is found that the model reproduces all the experimental tendencies.

INTRODUCTION

Interlocked materials are architectured materials from their very definition. Inspired from civil engineering, they are formed of blocs of a variety of geometries: tetrahedrons, cubes, osteomorphic pieces, etc. The interlocking property is obtained by the topology of the contacts and the application of compression loading conditions at the periphery of the assembly [1-2].

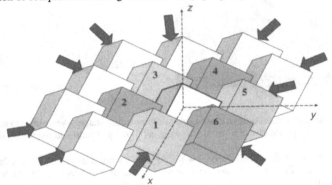

Figure 1: Example of an assembly of cubic elements.

A typical assembly of cubic shape elements is given in figure 1. Each cube is maintained by six neighbors. The vertical displacement along z < 0 of the central bloc is prohibited by cubes 1, 3 and 5. Similarly, the vertical displacement toward z > 0 direction is inhibited by the contacts with blocs 2, 4 and 6. Compression stress in the medium plane is needed in order to insure the contacts between all the blocs.

As shown by Dyskin *et al.* [3-4] for cubic blocs and by Autruffe *et al.* for osteomorphic blocs [5-6], the mechanical response in indentation is very sensitive to the quality of the contact between the blocs. Thus special attention should be paid in the contact laws when building up a numerical model devoted to such assemblies.

In this paper we have first realized assemblies of cubic blocs made of plaster. The effects of the surface roughness and the cube size have been investigated. Then, a discrete element model in which the interaction laws between two blocs are obtained by finite element simulations has been developed. Applications are then performed in the case of indentation loading and comparisons are made with the experimental observations.

INDENTATION EXPERIMENTS

A 270 x 280 mm assembly made of 30mm large plaster cubes has been realized and tested in indentation (see figure 2a). The number of cubes is 68. Each cube was mould in a dedicated frame which imposes five out of the six dimensions of the bloc. This induces uncertainties on one dimension of the cube that leads to gaps in the assembly. This effect can be partially removed by machining the sixth face of each cube.

Indentations were performed on the two realizations with several unloading stages. When the cubes are not machined, the mechanical response is found substantially lower than that of the machined assembly (figure 2b), due to the poor matching between the blocs. One could conclude that size precision strongly influences the mechanical strength of the structure.

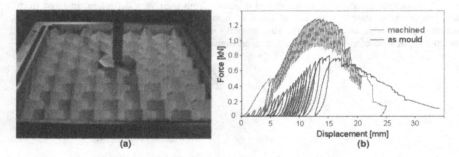

Figure 2: Experimental device used for indenting cubic shape assemblies (a) and indentation response obtained for the two types of cubes: mould and machined (b).

The effect of the cube size is now tested with a similar assembly (284 x 269mm) made of half size blocs. Figure 3a shows the assembly made of 264 blocs of 15mm large cubes. It is found that the smaller the blocs, the softer the mechanical response (see figure 3b). This could be explained by the increase of the number of surfaces in contact and consequently the accumulation of local relative displacements leading to a larger deflection.

144

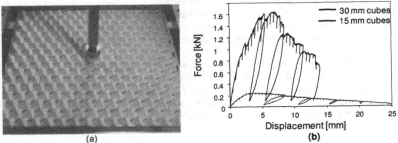

(a)   (b)

Figure 3: Indentation of 264 cubes of size 15mm (a) and comparison of the loading curves obtained for the two sizes investigated (b).

DISCRETE ELEMENT SIMULATIONS

The numerical model

A discrete element model is developed in order to better understand the experimental observations [7]. In this model, each cube is represented by its center coordinates and the triedron giving its orientation in space. Forces and momentum induced by the neighbor cubes are computed and sum up at the center. Then displacements and rotation of the cubes are evaluated in order to match the equilibrium state.

The key ingredient of the numerical model is the set of constitutive equations used to model the contact between two blocs. These have been obtained from finite element simulations of the interaction between two blocs. As shown in figure 4a, two quarters of blocs are meshed into 20 nodes elements. Compression force is first applied along X direction. Then, each of the six elementary motions is simulated by applying a relative displacement or rotation along the three possible directions and the corresponding reaction forces and momentum are recorded.

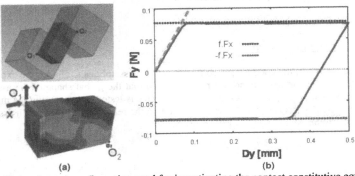

(a)   (b)

Figure 4: Finite element configuration used for investigating the contact constitutive equations (a) and example of the response obtained for a shear displacement applied along direction Y (b).

Since friction is taken into account in the FEM computations, a dissipative hysteretic curve is obtained (see figure 4) in the mechanical response. This is taken into account in the discrete element model through the introduction of an internal variable continuously updated with the history of the loading. In practice, the finite element response is approximated by analytical expressions.

In a first time, the code is validated in the case of a small assembly made of 19 blocs, i.e. two layers of cubes surrounding the indented central bloc. A typical loading curve is given in figure 5. It displays a hardening stage followed by a post-peak softening which leads to perforation when the force vanishes. When the friction coefficient is increased, the shape of the curve is modified. The peak force and displacement at perforation are both increased leading to a larger dissipated energy.

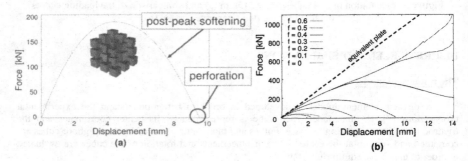

Figure 5: Discrete element simulations performed on 19 blocs (a)
and effect of the friction coefficient (b).

When the friction coefficient is above a critical value, the softening stage disappears and the loading curve transforms into a monotonic curve closer to the linear response we would obtain with a homogeneous plate of equivalent mass.

Comparisons with experiments

In a second step, the code is applied for bigger size assemblies similar to the experimental one. The three fold symmetry is however respected so that the global shape of the assembly is taken as hexagonal (see figure 6). Effect of the cube size is tested by comparing the mechanical response obtained with a set of 61 blocs of 30mm with the response obtained on an assembly of 169 blocs of 15mm.

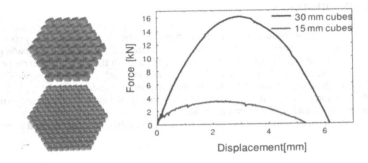

Figure 6: Influence of the cube size obtained by discrete element simulations

It is found that the peak force is eight times smaller for the smaller size (see figure 6). The experiments reported in figure 3b evidenced exactly the same trend with also a reduction coefficient of eight. The magnitude of the displacements and forces are also different. These discrepancies could be attributed to the boundary conditions and more specifically to the magnitude of the compression boundary conditions. Indeed, in the experiments, the compression force was not precisely mastered. It was imposed by a set of screws located in the outer rigid frame. Numerically, it is found that the prestress strongly influences the magnitude of the indentation force. Thus, only qualitative agreement can be obtained at this point.

The effect of the compression stress is now investigated in figure 7 where both experiments and simulations are presented together. When the torque applied to the frame screws is twice higher, the peak force is increase by a factor 1.3. In the discrete element model, the peak force is multiplied by the same factor when the compression force is doubled.

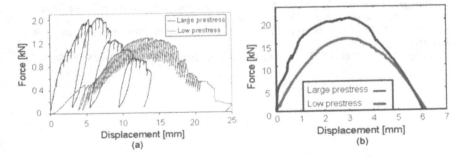

Figure 7: Influence of the compression loading conditions measured experimentally (a) and simulated by the discrete element model (b).

Once again, the discrete model matches the experimental observations.

CONCLUDING REMARKS

A discrete element code has been developed in order to reproduce the experimental behavior of large assemblies of interlocked blocs submitted to indentation loading. The numerical tool has been able to reproduce the main tendencies in the case of cubic shape blocs. The tool can now be used to quantify the effect of parameters such as the friction coefficient or the magnitude of the compression stress. In a forthcoming study, the code will be applied to treat the case of osteomorphic blocs. It could also be used to investigate the case of multimaterials, i.e. when the assembly is made of different materials.

Finally, the numerical tool will be useful to check the behavior of interlocked structure in term of damage.

ACKNOWLEDGMENTS

The authors would like to thank Prof. Y.Estrin for triggering our interest in these new materials, via many stimulating discussions.

REFERENCES

1. A.V. Dyskin, Y. Estrin, E. Pasternak, H.C. Khor and A.J. Kanel-Belov, *Adv. Eng. Mater.* **5**, 116 (2003).
2. A.V. Dyskin, Y. Estrin, A.J. Kanel-Belov and E. Pasternak, *Phil. Mag. Letters* **83**, 197 (2003).
3. S. Schaare, A.V. Dyskin, Y. Estrin, S. Arndt, E. Pasternak and A.J. Kanel-Belov, *Int. J. of Eng. Sci.* **46**, 1228 (2008).
4. Y. Estrin, A.V. Dyskin, E. Pasternak, S. Schaare, S. Stanchits and A.J. Kanel-Belov, *Scripta Mater.* **50**, 291 (2004).
5. C. Brugger, A. Autruffe, Y. Bréchet, M. Fivel, and R. Dendievel, *in Proc. of 18ème Congrès Français de Mécanique,* Grenoble, France, August 27-31 (2007).
6. A. Autruffe, F. Pelloux, C. Brugger, P. Duval, Y. Bréchet and M. Fivel, *Adv. Eng. Mater.* **9-8**, 664 (2007).
7. C. Brugger, Y. Bréchet and M. Fivel, *Advanced Materials Research* **47-50**, 125 (2008).

Mater. Res. Soc. Symp. Proc. Vol. 1188 © 2009 Materials Research Society   1188-LL05-04

Superelasticity, Shape Memory and Stability of Nitinol Honeycombs Under In-Plane Compression

John A. Shaw[1], Petros A. Michailidis[1], Nicolas Triantafyllidis[1], and David S. Grummon[2]

[1]University of Michigan, Aerospace Engineering, Ann Arbor, Michigan, 48109-2140
[2]Michigan State Univ., Chemical Engin. & Materials Science, East Lansing, Michigan, 48824

ABSTRACT

Low density Nitinol shape memory alloy honeycombs were fabricated using a new Nb-based brazing method [1], which demonstrated enhanced shape memory and superelastic properties under in-plane compression [2]. Adaptive, light-weight cellular structures present interesting possibilities for design of new architectures and novel applications. This paper presents an overview of ongoing work to address the multi-scale stability of superelastic, thin-walled, SMA honeycombs and the need for design and simulation tools.

BACKGROUND

Shape memory alloys (SMAs) are a material class that exhibit the shape memory effect and superelasticity [3], two strain recovery phenomena occurring with changes in temperature and/or stress, that can enable novel adaptive and energy absorption applications. Popular commercial SMAs are NiTi-based (near equiatomic Nitinol, or NiTiX alloys), which have the robust strain recovery and structural properties as polycrystals. On the other hand, cellular metals, made of conventional materials, like aluminum, are desirable in applications where low density, high stiffness, and energy absorption are needed [4]. Papka and Kyriakides [5] performed interesting in-plane crushing experiments of thin-walled aluminum honeycombs, where structures exhibited an initially stiff response, followed by a load plateau with localized row-by-row elasto-plastic collapse and permanent deformation, and then stiffening from internal cell contact. We were inspired to explore whether adaptive honeycomb structures, using SMAs, could be made that recover deformation after load removal (superelasticity) or upon heating (shape memory).

Historically, joining Nitinol to itself has been difficult, usually requiring engineers to use mechanical fasters or adhesives [6, 7]. Metallurgical bonding required special welding techniques and conventional brazing usually resulted in weak properties. A new metallurgical bonding method for NiTi, however, was discovered [1] that has good strength, ductility, corrosion resistance, and biocompatibility [8]. Figure 1 shows photographs and micrographs of Nitinol tubes that were brazed together using Niobium as a melting point suppressant, where a quasi-eutectic resulted in aggressive wetting and formation of a classical lamellar microstructure

Figure 1: Niobium-based brazing of Nitinol tubes.

of NiTi and bcc-Nb. This enabled the construction of the first NiTi honeycomb specimens (near 5 % dense) with useful adaptive properties. Hexagonal and lens-like cell geometries were produced by shape setting NiTi strips into corrugations, bonding them together at high temperature using Nb, and then heat-treating near 500 °C. In this way it is now possible to build-up various architectures from wrought materials. Compressive isothermal experiments on fabricated specimens exhibited superelasticity with over macroscopic 50 % strain recovery [2] as shown in Fig. 2a, where a hexagonal specimen was subjected to progressively larger displacement load-unload cycles. The homogenized (or macroscopic) compressive stress and strain are reported as the ratios of compressive force to initial footprint area (F/A) and compressive displacement to initial height (δ/H). Non-isothermal shape memory experiments on a wavy-corrugated specimen demonstrated stress-free recovery of macroscopic strains approaching 20 % [9] as shown in Fig. 2b, where the specimen was subjected to a shape memory cycle (stress-free cooling, isothermal load-unload, and stress-free heating) and a high temperature superelastic cycle.

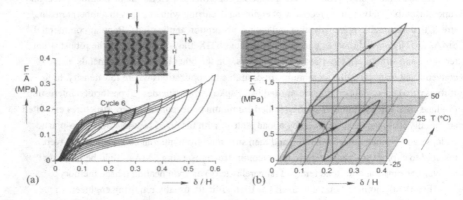

Figure 2: (a) Compressive experiment on superelastic Nitinol honeycomb (room temperature) [2]. (b) Compressive shape memory, then superelastic, experiment on Nitinol corrugation [9].

SMAs have the largest mechanical energy density of all known adaptive materials by one or two orders of magnitude [10] due in large part to the large stress capability. A low-density SMA

150

cellular architecture has several advantages over its monolithic forms. The recoverable tensile strain recovery of monolithic Nitinol is about 6 % for low-cycle operation and 2.5 % for high cycle operation. These limits can be "leveraged" by exploiting bending of thin walls in an open cell topology. In addition, heat transfer limitations make SMA actuators thermally sluggish and cause hypersensitive rate-dependencies [11]. Since the thermal time constant scales with the material's volume to surface ratio, it can be reduced by adopting a low-density architecture. Built-up architectures now offer the design flexibility to trade off force for displacement to achieve adaptive properties in hitherto vacant force-displacement regimes. The new design game is to develop architectures with sufficient stiffness and strength, while keeping local fiber strains within recovery limits. The maximum bending strain for a given global compressive strain scales with the ligament thickness to length ratio t/l, the plateau stress scales as $(t/l)^2$, and the initial structural stiffness scales as $(t/l)^3$ [12]. Since the ligament thickness ratio can be made about 10× larger for Nitinol than for aluminum alloys to avoid permanent strain for a given topology, the Nitinol structure will have greater strength and stiffness by about 100× and 1000×, respectively.

MODELING & SIMULATION

Considering the potential design opportunities, we developed a finite element simulation tool to explore the superelastic response of a honeycomb structure, investigate instabilities, and study the influence of material properties and geometry. A hexagonal honeycomb is shown in its reference configuration in Fig. 3a. The cell walls have thickness t (except for $2t$ along the horizontal bonds) and length l. The periodic unit cell shown was used in Bloch wave calculations to assess stability with respect to periodic perturbations of all wavelengths. The cell wall thickness ratio was $t/l = 1/30$ as in the specimen of Fig. 2a. During in-plane compression along the X_2-axis, the cell walls deform through bending and minor membrane deformation, so each ligament was idealized as small strain, nonlinear beams (arbitrarily large displacements and rotations) in an incremental finite element analysis. The local fiber stress-strain ($\sigma - \varepsilon$) law was a single internal variable (martensite fraction), piecewise-linear, hysteretic superelastic model as shown in Fig. 3b. The influence of different material parameters were investigated, including transformation strains (β^+, β^- in tension and compression, respectively), nucleation strains (ε_n^+, ε_n^-), transformation tangent moduli (E_t^+, E_t^-), and stress hysteresis ($\Delta\sigma^+$, $\Delta\sigma^-$). See [13] for more details and an extensive parameter study.

As a simplified example, Fig. 4 shows a finite element simulation of the principal branch (periodic deformation of the unit cell) under displacement controlled loading of a perfect (infinite) honeycomb using a nonlinear elastic material. The mechanical response shown in Fig. 4a exhibits an initially stiff response, then a load "plateau" due to geometric and material softening, followed by a gentle rise as the material stiffens and transformation saturates. The $\wedge$ and $\vee$ denote limit

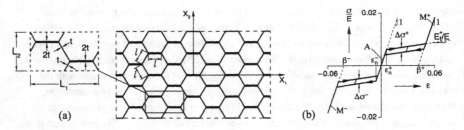

Figure 3: (a) Honeycomb geometry and unit cell, (b) constitutive model.

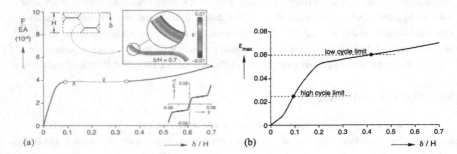

Figure 4: (a) Finite element simulation of the dimensionless compressive force-displacement response for a perfect, infinite honeycomb with "nonlinear elastic" material. (b) Maximum local strain (ε_{max}) versus global compressive strain (δ/H) during the simulation.

loads, and the open circles denote critical points where the stability of the structure changes according to the Bloch wave analysis. The critical points were found to be infinite wavelength modes for this load orientation, while intermediate equilibrium points are a continuum of bifurcation points with multiple finite wavelength eigenmodes. The thin line denotes the unstable segment of the path, so the response is initially stable, looses stability, and then regains stability during monotonic loading. The inset shows strain contours of the slanted cell ligament at $\delta/H = 0.7$ with worst case strains in the corners (bending dominated). The evolving maximum local strain (ε_{max}) is plotted against the global strain (δ/H) in Fig. 4b. The dotted lines show that high cycle (0.025) and low cycle (0.06) limits are reached at global strains near $\delta/H = 0.1$ and $\delta/H = 0.42$, respectively, which demonstrates the kinematic strain amplification of the thin walled structure. Such an analysis provides structural limits to avoid permanent deformation, with the caveat that less symmetric configurations occur wherever the principal branch is unstable.

Figure 5 shows a comparison of responses of finite and infinite (periodic) honeycombs using a hysteretic material (Fig. 3b). The finite honeycomb is compressed by frictionless rigid platens at its top-bottom edges while unloaded along the left-right edges. Figure 5a shows the mechanical

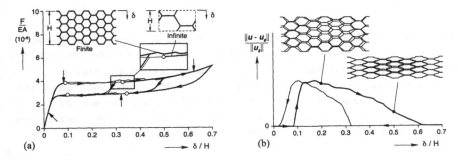

Figure 5: (a) Finite element simulations of perfect finite and infinite honeycombs using hysteretic material. (b) Deviation of finite to periodic (infinite) honeycomb configurations.

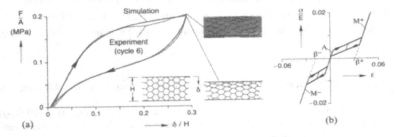

Figure 6: Simulation of experiment: (a) mechanical response and configurations using the actual imperfect geometry and frictional platens ($\mu = 0.3$), (b) asymmetric constitutive model.

responses with unloading initiated at $\delta/H = 0.3, 0.5$, and 0.7. The responses of the infinite and finite honeycombs are rather similar, except the infinite case has unstable segments during loading and unloading while the finite honeycomb is always stable with minor oscillations about the perfect case. The arrows in the figure show where the finite honeycomb response diverges then converges to the infinite one during loading and unloading. The differences are more apparent in Fig. 5b where a suitable norm (see [13]) was selected to measure the deviation of the displacement fields between the finite (u) and periodic ($u_\#$) cases. Selected configurations show the envelope of the finite case becoming trapezoidal by drawing inwards along the bottom edge. Adding random imperfections to the finite honeycomb (not shown) amplifies this effect (see [13]). Lastly, to verify reasonableness of the approach a simulation was performed (see Fig. 6) using a more realistic material law with texture effects (tension-compression asymmetry), the actual imperfect geometry (obtained from a high resolution photograph of the actual specimen), and some friction at the platens to capture cycle 6 of Fig. 2a. The results show satisfactory agreement (with minor deviations) with both the measured response and deformed configurations.

CONCLUSIONS

The realization of SMA honeycombs that exhibit an amplified form of superelasticity and shape memory provides an opportunity to design new adaptive structures tailored to the users needs by choice of internal cellular architecture. This paper provided an overview of the enhanced superelastic in-plane compression behavior of fabricated Nitinol honeycombs and presented finite element simulations capable of capturing multi-scale instabilities arising from geometric and material nonlinearities. Work is ongoing to explore alternative cellular forms and to optimize their design for specific energy absorption and thermally-actuated applications.

ACKNOWLEDGMENTS

Financial support is gratefully acknowledged from the National Science Foundation (CMS0409084, manager Dr. K. Chong), the U.S. Air Force Office of Scientific Research (FA9550-08-1-0313, manager Dr. L. Lee), and Nitinol Devices & Components (Dr. A. Pelton).

REFERENCES

[1] D. Grummon, J. Shaw, and J. Foltz, Materials Science and Engineering, A **438-440**, 1113 (2006).
[2] J. A. Shaw, D. S. Grummon, and J. Foltz, Smart Materials and Structures **16**, S170 (2007).
[3] K. Otsuka and C. M. Wayman, editors, *Shape Memory Materials*, Cambridge University Press, Cambridge, UK, 1998.
[4] M. Ashby et al., *Metals Foams: A Design Guide*, Butterworth–Heinemann, Boston, MA, 1st edition, 2000.
[5] S. Papka and S. Kyriakides, Acta Materialia **46**, 2765 (1998).
[6] M. R. Hassan, F. L. Scarpa, and N. A. Mohamed, Shape memory alloys honeycomb: design and properties, in *Proceedings of SPIE: Smart Structures and Materials 2004*, volume 5387, pages 557–564, SPIE, 2004.
[7] Y. Okabe, S. Minakuchi, N. Shiraishi, K. Murakami, and N. Takeda, Advanced Composite Materials **17**, 41 (2008).
[8] D. S. Grummon, J. A. Shaw, and K.-B. Low, Reactive eutectic brazing of nitinol using niobium, in *Proceedings of ICOMAT08*, Sante Fe, NM, United States, 2008.
[9] J. A. Shaw et al., Shape memory alloy honeycombs: Experiments and simulation, in *Proceedings of the AIAA, ASME, ASCE, AHS, ASC Structures, Structural Dynamics and Materials Conference*, volume 1, pages 428 – 436, Waikiki, HI, United States, 2007.
[10] P. Krulevitch et al., Journal of Microelectromechanical Systems **5**, 270 (1996).
[11] M. A. Iadicola and J. A. Shaw, International Journal of Plasticity **20**, 577 (2004).
[12] L. J. Gibson and M. F. Ashby, *Cellular Solids: Structure and Properties*, Cambridge Solid State Series, Cambridge University Press, Cambridge, UK, 2nd edition, 1997.
[13] P. Michailidis, N. Triantafyllidis, J. Shaw, and D. Grummon, International Journal of Solids and Structures **46**, 2724 (2009).

Multifunctional Materials

Mater. Res. Soc. Symp. Proc. Vol. 1188 © 2009 Materials Research Society   1188-LL06-06

Multifunctional Cellular Materials

Michel de Gliniasty[1] and Régis Bouchet[2]
[1] ONERA, BP 72, 29 avenue de la division Leclerc
FR-92322 Chatillon Cedex
E-mail : gliniasty@onera.fr
[2] ETOP International

ABSTRACT

In many industrial fields, structural materials play a key role in the increase of performance, but new requirements in terms of energy saving, safety, materials economy... lead to more stringent requirements on materials properties.

The two usual strategies –microstructure optimisation and shape optimisation-, which act at two different scales, the micrometer scale and above the centimetre scale, become less and less efficient with this new strong demand for multi-functional properties. The largely unexplored millimetre scale, domain of the so called "structural materials", is a possible answer. Structural materials benefit of an extra degree of freedom well suited for multi-functionality: they allow using combination of materials from different classes, allow geometrical optimisation, and can be naturally integrated in structures such as sandwiches and various stiffened plate geometries.

The price to pay for this extra-richness is the extraordinary wide variety of potential solutions to investigate for a given problem. Hence modelling plays a crucial role for selecting and optimising such innovative materials.

This paper is an overview of a project, named MAPO ("Materiaux Poreux"), aiming at designing high-temperature materials with acoustical and structural properties.

INTRODUCTION

MAPO is a common project between ONERA, the French aerospace lab, and CNRS the French national centre for research. The objective of this project, launched in 2003 and of 5 years duration, was to explore the "materials by design" approach. Most of the work has been carried out through PhD thesis [1].

The concept of designing a special material for a specific application is not original. At the microstructure scale, the excellent paper by S. Naka and T. Khan [2] is a key contribution in the domain and at the mesostructure scale composite materials are very good examples.

In both cases designing a new material was imposed by the required multi-functionality. In the first example high creep tensile strength and good ductility were the –usually contradictory- required properties and the authors attempted to create a γ–γ' type microstructure in several new alloy systems. Their attempt was not successful as our knowledge of the relationship between microstructure and macroscopic properties is far from being satisfactory for that purpose. In the second example the resulting materials were efficient enough to lead to applications, showing that intermediate scale parameters are easier to handle.

Within the MAPO project it was decided to undertake work on "structured" materials, which means that the degree of freedom would be given by the geometry at a millimetre scale. This is by no means new, but it is usually applied at scales above centimetres.

MATERIALS REQUIREMENTS AND CHOICE

Since ONERA is an aerospace lab, the chosen functionalities were thermo-mechanical strength and acoustic damping capabilities for linings of internal surfaces of aeroengines nacelles; of course light weight was also a key consideration.

The operating temperature (several hundreds Celsius) ruled out polymeric materials and structural properties excluded ceramics. Hence the constitutive material had to be metallic. The required acoustic properties imposed an open porosity and as MAPO was conceived as an exploratory project, two types of generic geometries were selected: microtubes and metallic hollow spheres. Finally the second geometry was selected in a preliminary approach. It is interesting to note here that the choice has been dictated also because of the cristallographic analogy: hollow sphere structures allow fcc, cc, simple cubic and hcp stacks preferred by metallurgists. But the difficulty of modelling such stacks has been forgotten: the demonstration of Kepler's conjoncture on the most compact arrangement of equal spheres, proposed in 1611, has been demonstrated in 1998 with 250 pages of fastidious mathematics and its latest simplification has been published in 2007 [3]!

As mentioned above, the objective of the project MAPO was to optimise the material for the proposed application by varying the main parameters of the cellular material, e.g. the diameter and thickness of the spheres. But there are many other parameters: the constitutive material, the type of stacking (regular or randomly packed) and the process for joining the spheres.

Unfortunately, during the five years of the project, the materials changed and the different results are not always comparable. At the beginning of the project regularly packed nickel hollow sphere structures, provided by the French company ATECA, with a Ni-B brazing process for assembling the spheres, were mainly used. Later on 314L stainless steel hollow sphere structures, provided by the Austrian company PLANSEE, were also used and all the stacks were randomly packed (gravity driven). Table 1 displays the main characteristics of the studied materials.

Table.1.Materials

Source	Material	Density	Diameter	Thickness	Wall porosity	Young's modulus	Yield stress
Ateca	Ni&NiB	1000g/l	2.8mm	120μm	Dense	700±100MPa	3.7±0.2MPa
Plansee	314L	800g/l	2.8mm	90μm	Dense	2000±200MPa	6.8±0.1MPa
Plansee	314L	600g/l	2.8mm	65μm	Dense	1600±50MPa	5.4±0.1MPa
Plansee	314L	400g/l	2.8mm	45μm	Dense	1000±50MPa	3.2±0.05MPa
Plansee	314L	800g/l	1.7mm	45μm	dense	2000±200MPa	8±0.5MPa

PLANSEE materials were more reproducible, and as reproducibility became a strong demand, especially because of the mechanical properties, these materials were mostly used for the mechanical studies.

SELECTED APPROACHES

Two different approaches are possible. The first one relies on direct experimental characterization of the aimed properties; several experiments are necessary, with many different materials in order to derive phenomenological macroscopic laws. This approach will be referred, hereafter, as the descriptive method.

The second approach is more predictive: it consists in solving the basic equations of the physical problem for a representative cell of the structure, i.e. at the mesoscopic scale (for instance the representative figure of a cc stack would be a sphere surrounded by eight eighths of spheres). This can be done by numerical calculations when adding periodic boundary conditions. Using periodic homogenization methods, the properties of an equivalent homogeneous medium are then calculated and with several calculations it is possible to calibrate general laws. This descriptive method is easy when it is possible to define without ambiguity the representative cell, as in the case of regular stacks. But when the spheres are randomly packed it is more difficult; A. Fallet [1] has proposed an interesting method to solve that problem (see next section).

In both approaches, a continuous medium physical model is necessary; the chosen model for mechanical behaviour is Hashin's [4] and for acoustic properties the model is derived from Kirschoff's works.

The second approach being more adequate for optimising a material, most of the work within MAPO was dedicated to predictive methods and especially to *a priori* modelling of structured materials.

MECHANICAL PROPERTIES

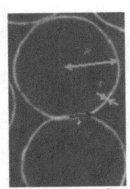

Fig.1 Mesoscopic scale parameters

Some previous investigations using descriptive methods to calculate mechanical properties of hollow spheres structures exist (see [5], [6] for example) but the amount of experimental data is small and the domains of validity of the models are limited. From a mechanical point of view, the pertinent parameters are the relative density ρ_r, the radius R and thickness t of the spheres, the radius a of the contacts and the coordination number Z (see Fig.1).

The results usually describe the relative Young's modulus E/E_s and the relative yield strength σ_y/σ_{ys} as functions of the ratios t/R and a/R (see Gasser [1] for example). For regular stacks, the relative density is mostly a function of t/R.

In the case of randomly packed spheres, A. Fallet's method to build a predictive model is based on the experimental characterization of the stacks using X-ray tomography. The true values of t/R and a/R as well as the coordination number of the materials are then available. Moreover, he has shown that for a given manufacturing process (e.g. the PLANSEE process), there is a

159

relationship between the relative density and the coordination number. A predictive method is then achievable for different materials manufactured by the same process.

The mechanical properties are then computed through different models: FEM (Zebulon software [7]) to derive the interaction laws between two spheres as a function of the constitutive material properties and of the structural parameters R, t and a; numerical generation of random stacks of a representative number of spheres through the method proposed by F. Mamoud in his PhD dissertation [1], and finally smooth-DEM (dp3D software [8]) to compute the macroscopic mechanical properties. The computations are limited to 3% strain because of the DEM method, but it is sufficient to get the Young's modulus E and the yield strength σ_y or $\sigma_{0.2}$.

The derivation of scaling laws with the structural properties is then straightforward; once the relative thickness t/R is given, the manufacturing process determines the other parameters. The problem is that if t/R is relatively constant in the stacking, it is not the case for a/R and an equivalent quantity has to be defined. As the important parameter is obviously the surface of the contacts between the spheres, a new structural parameter is defined:

$$a_{eq}/R = [1/n\Sigma \ (a_i/R)^2]^{1/2}$$

the sum being made on all the n contacts of the sample.

Fig.2 gives the relative modulus and yield strength results as functions of the relative density, parameterised with t/R and a_{eq}/R.

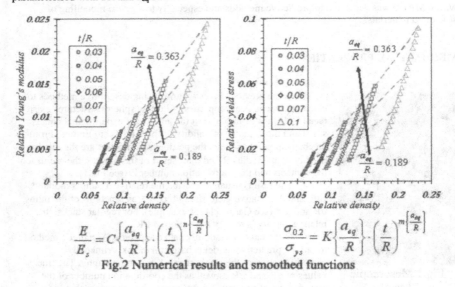

$$\frac{E}{E_s} = C\left\{\frac{a_{eq}}{R}\right\}^n \cdot \left(\frac{t}{R}\right)^{\left\{\frac{a_{eq}}{R}\right\}} \qquad \frac{\sigma_{0.2}}{\sigma_{ys}} = K\left\{\frac{a_{eq}}{R}\right\}^m \cdot \left(\frac{t}{R}\right)^{\left\{\frac{a_{eq}}{R}\right\}}$$

Fig.2 Numerical results and smoothed functions

At this stage it is worth noting that a given elastic modulus may be obtained within a wide range of densities, provided that the equivalent contact area is adapted. An interesting feature of the work done by A. Fallet is the successful comparison with experimental results coupled to a characterization of the behaviour of the material during compression at the mesoscale by in situ X-ray tomography.

As the tomography provides the equivalent contact area of the samples, the comparison is easy. Simulations give good results concerning the stress-strain curves and especially for the stress plateau, but they overestimate the compression modulus (cf Fig.3). This could be attributed to the unavoidable variation of the radius and thickness of the spheres.

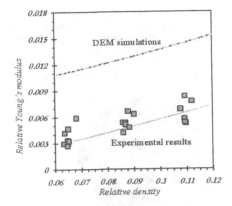

Fig.3 Numerical results vs experiment

ACOUSTIC PROPERTIES

The problem is the following: given a porous material, how to calculate its absorption as a function of the frequency of incident acoustic waves. The basic descriptive model has been derived by Kirschoff in the case of normal acoustic waves and for cylindrical tubes. It leads to the expression of an effective density ρ_{eff} and an effective compressibility χ_{eff}, depending on the frequency and allowing the calculation of the absorption curve.

For more complex porous materials, most descriptive approaches are based on Biot's [9] works. The model used by S. Gasser in his thesis is the called Pride & Lafarge model [10], [11]. It relies on eight parameters: open porosity ϕ, tortuosity α, a characteristic length Λ, two shape factors C_p and p for the pores, characteristics of the viscous properties and their equivalent Λ', C'_p and p' characteristics of the thermal properties.

All these models, which suppose the existence of a representative equivalent medium, require some assumptions: the solid matrix must be rigid and isothermal, which is acceptable for the hollow sphere stacks, the material must be isotropic (elastically and thermally) at a macroscopic scale and the pores must not be connected. Obviously regular stacks of spheres are not isotropic, but they exhibit many symmetries that compensate (see the work of N. Auffray [1] in his PhD

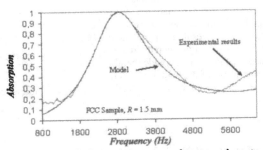

Fig.4 Biot's model compared to experiment

dissertation about symmetries in cellular materials), and their pores are connected. Nevertheless, when adequately calibrated the model gives very good results (see Fig.4).

S. Gasser has proposed an interesting predictive approach based on a finite elements modelling of the elementary cell of a fcc stack. Starting from locally expressed continuity, momentum and energy (including heat) equations, along with the hypothesis of homogeneity of the pressure and appropriate boundary conditions, he has developed appropriate finite elements softwares. Applying a two-scales homogeneization method, he is then able to calculate the effective density and compressibility and produce absorption curves at different thicknesses L of the material, without any adjustment.

Fig.5 shows the elementary cell and its mesh, while Fig.6 gives the result compared to experiment in the case of 2mm spheres in a 20mm thick material.

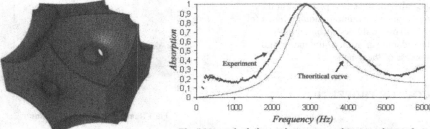

Fig.5 Elementary cell   Fig.6 Numerical absorption compared to experimental one

As expected the absorption curve is only dependant on the sphere radius R and on the sample thickness L. But it also appears that the relative contact parameter a/R, can influence the results. Without any information on typical variations, S. Gasser has deliberately discarded this parameter, which fortunately is probably of second order. The agreement between calculations and experiments being acceptable, it has been possible to describe the theoretical absorption as a function of the radius of the sphere for different thicknesses of samples (Fig.7).

It is worth noting that there is a radius for which absorption is maximum: the explanation is that for small radii the permeability is low and the sound waves do not penetrate easily the pores although when the radius is large, the pores are so wide that the relative size of the boundary layer, where dissipation occurs, is small.

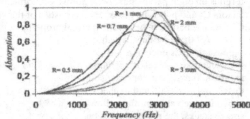

Fig. 7 Absorption for different radii and for a 2cm thickness

OPTIMISATION

Although the results are not complete, one can imagine the optimisation procedure. Given an acoustic spectrum and a maximum thickness L imposed by the engine geometry, the best radii R can be derived (the optimum is usually flat, even in the case of a peaked noise spectrum, see [12]). The required mechanical properties, E and σ_y, defining the stiffness and the acceptable strain, allow deriving couples of relative thicknesses t/R and relative contact sizes a_{eq}/R fulfilling the conditions. The largest a_{eq}/R leading to the lowest density, hence the lowest mass; the next step is, of course, to check that the manufacturing process is feasible. Iterations could also be made on the couple L (part thickness) and density ρ of the material.

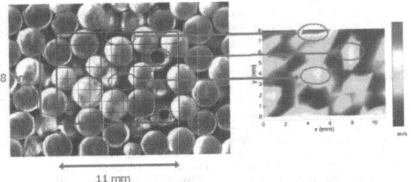

But this is a very theoretical procedure; besides the fact that the different materials studied during the project are not identical, the models as well as the experimental results have been obtained without any flow and with normally impinging acoustic waves.

Recent experimental results have been obtained by using a new method (see the works of F. Simon [13]) with flows up to Mach 0.4. Fig. 8 describes the experimental arrangement.

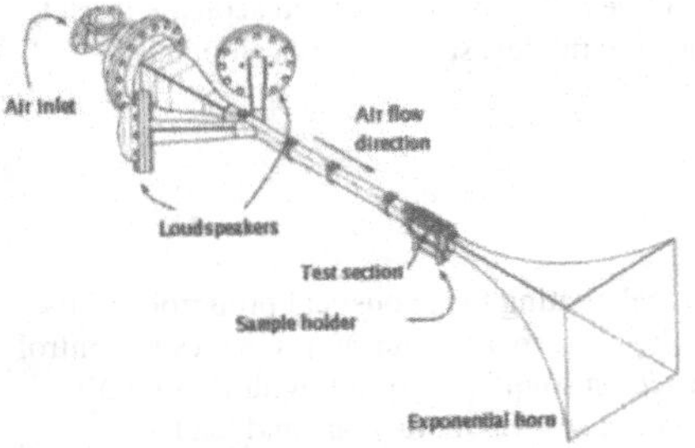

Fig.8 Experimental arrangement for acoustic tests with flow

The measurement technique is Laser Doppler Velocimetry (2D at the moment) very close to the sample surface. With the help of advanced techniques of signal processing the experiment gives access to the acoustic velocity and to the thickness of the acoustic boundary layer. Fig.9 shows the results obtained at M 0.1 with the ATECA material (see Table 1).The precision of the method being very good, one can see that the normal acoustic velocity is maximum just above the holes in the material.

Fig-9 Measured normal acoustic velocity at surface level, with M0.1 flow

Knowing the acoustic velocity, it is possible to calculate the parameters of Kirchoff's model and thereby this new type of experiments may be used in the optimization process. But this has not been done within the project.

Other experimental results demonstrate that the hollow sphere materials are globally less efficient with flow in comparison with usual sandwiches with honeycomb materials. The proposed explanation is that the connected porosity of the hollow spheres let the acoustic waves find a shortest way to exit. They are nevertheless rather insensitive to the Mach number up to 0.4.

But the most significant result obtained with the LDV experiment is the demonstration that above 120dBs the linearity hypothesis is no longer valid (see M. Lavieille [1]); the damping process becomes different with the onset of small vortices. This means that the acoustic models used in the project are no longer valid and that some new models should be developed.

PERSPECTIVES AND CONCLUSIONS

It should be possible to unify the results by recalculating the acoustical properties of the random stacks from PLANSEE, and try to develop elaborate manufacturing processes to control the parameter a_{eq}/R of section 3 and check its influence on sound absorption with F. Simon's method...Such a work would require several more thesis, several more years and lead to a limited domain of application, particularly because of the linearity assumption. This was not intended and the project stopped as in the case of hollow spheres materials work.

Besides the development of new techniques, of new models, of interesting couplings...the main lesson of MAPO is that a simple or elegant geometry is not the correct approach for choosing a test structured material. A good test material should be manufactured at demand, should be reproducible, and its parameters should be easily varied.

Enriched by this experience, ONERA is now working on other structured materials, such as assemblies of tubes which, albeit two dimensional and hence non isotropic, will fulfil the above mentioned properties.

It is also worth noting that MAPO has been a unique opportunity to mix researchers from different disciplines, different laboratories and to teach young scientists that the future lies in multi-disciplinary research.

REFERENCES

[1] More than seven thesis have been partially or totally supported by the project. The following are directly connected to this paper:
S. Gasser, PhD. Thesis, Institut National Polytechnique de Grenoble, 2003
A. Fallet, PhD. Thesis, Institut National Polytechnique de Grenoble, 2008
F. Mamoud, PhD. Thesis, Institut National Polytechnique de Grenoble, 2007
N. Auffray, PhD. Thesis, Institut National Polytechnique de Grenoble, 2008
M. Lavieille, PhD. Thesis, Institut Supérieur de l'Aéronautique et de l'Espace, Toulouse, 2007
[2] S. Naka and T. Khan in *Intermetallic Compounds*, vol. 3, Eds J. H. Westbrook and R. L. Fleisher, (John Wiley and Sons, 2002), p.841
[3] C. Marchal ONERA Report N° 2007-01 (2007)
[4] Z. Hashin, J. App. Mech. **29**, 143-150 (1962)
[5] W.S. Sanders, PhD. Thesis, Massachusetts Institute of Technology, 2002
[6] I. Taguchi, Journal of Mechanics of Materials and Structures, **2** (3), 529-555 (2007)

[7] ZéBuLoN, www.numerics.com
[8] C. L. Martin, habilitation to direct PhDs, Institut National Polytechnique de Grenoble, 2005
[9] M.A. Biot, J. Acoust. Soc. Am. **28** (2), 168-191 (1956)
[10] S.R. Pride et al., Phys. Rev.B **47**, 4964-7978 (1993)
[11] D. Lafarge, J. Acoust. Soc. Am. **102** (4), 1995-2006 (1997)
[12] S. Gasser, Y. Brechet, F. Paun, Advanced Engineering Materials **6**, 97 (2004)
[13] A. Minotti, F. Simon, F. Gantie, Aerospace Science and Technology, **12** (5), 398-407 (2008)

Mater. Res. Soc. Symp. Proc. Vol. 1188 © 2009 Materials Research Society

Design of Nano-Composites for Ultra-High Strengths and Radiation Damage Tolerance

A. Misra[1], X. Zhang[2], M. J. Demkowicz[3], R. G. Hoagland[1] and M. Nastasi[1]
[1] Los Alamos National Laboratory, MS K771, Los Alamos, NM 87545
[2] Texas A&M University, College Station, TX
[3] MIT, Cambridge, MA

ABSTRACT

The combination of high strength and high radiation damage tolerance in nanolaminate composites can be achieved when the individual layers in these composites are only a few nanometers thick and therefore these materials contain a large volume fraction associated with interfaces. These interfaces act both as obstacles to slip, as well as sinks for radiation-induced defects. The morphological and phase stabilities of these nano-composites under ion irradiation are explored as a function of layer thickness, temperature and interface structure. Using results on model systems such as Cu-Nb, we highlight the critical role of the atomic structure of the incoherent interfaces that exhibit multiple states with nearly degenerate energies in acting as sinks for radiation-induced point defects. Reduced radiation damage also leads to a reduction in the irradiation hardening, particularly at layer thickness of approximately 5 nm and below. The strategies for design of radiation-tolerant structural materials based on the knowledge gained from this work will be discussed.

KEYWORDS: Irradiation damage, dislocations, multilayers, interfaces

INTRODUCTION

The performance of structural materials in extreme environments of irradiation and temperature must be significantly improved to extend the reliability, lifetime, and efficiency of future nuclear reactors [1]. In reactor environments, damage introduced in the form of radiation-induced defects and helium results in embrittlement and dimensional instability. Therefore, the ability to control radiation-induced point defects and helium bubble nucleation and growth is a crucial first step to improving the mechanical properties of irradiated metals. This calls for novel approaches in designing structural materials that resist radiation damage while maintaining high strength and toughness. Using nanolayered metallic composites as model systems, we highlight how tailoring the length scales (layer thickness) and interface properties can provide insight into ways of designing radiation damage tolerant structural materials.

The experiments described here were conducted on sputter deposited Cu-Nb multilayers and single layer Cu and Nb samples using He^+ ion implantation to cover a broad range of irradiation conditions and layer thicknesses as described below [2-5] (throughout this article, multilayers are designated by the individual layer thickness or one-half of the bilayer period. Thus, 2 nm multilayer refers to a sample with a bilayer period of 4 nm with 2 nm Cu and 2 nm Nb layers). Cu and Nb have a positive heat of mixing, very limited solid solubility, and no tendency to from intermetallic compounds. Thus, the interfaces between Cu and Nb layers are very distinct with little evidence of mixing. But irradiation and/or temperature can lead to morphological instabilities. The spheroidization of layered structures upon thermal annealing is well known. However, our earlier work on vacuum annealing of Cu-Nb nanolayered composites did reveal remarkable thermal stability in these materials with no evidence of spheroidization up to 800 °C [6]. This observation formed the basis to explore the stability of nanolayered composite materials under irradiation conditions. Similarly, radiation may destroy the geometrical arrangement of the layered structure because the flow of radiation-induced point defects into an interface changes its atomic configuration, perturbing its shape from flat to meandering. Such undulations provide an opportunity for adjacent nanometer-spaced interfaces to react and pinch off. Because the stability of the layered structure relies upon the planarity of the interfaces, such reactions can destabilize the structure and trigger the onset of recrystallization. There is a competition between restoring forces acting to flatten the interface and the undulations induced by rapid adsorption of defects. Chemical and morphological stabilities of the interfaces are necessary factors in the design of nanolayered composites where interfaces are to act as sinks of radiation-induced defects.

RESULTS AND DISCUSSION

The results presented below are organized to highlight the roles of interfaces in (i) suppression of helium bubble nucleation, (ii) retardation of helium bubble growth at elevated temperatures, (iii) reducing irradiation hardening, and (iv) producing high strength materials.

Delayed Nucleation of Bubbles in Nanolayered Composites

The equilibrium solid solubility of helium in metals such as Cu and Nb is extremely low, on the order of parts per billion or smaller. For helium concentrations above the solubility limit, helium tends to cluster and precipitate in the form of bubbles. However, the solubility of helium is expected to be higher at grain boundaries and interphase interfaces. As an upper bound, if we estimate that one monolayer of helium can be in solution at an interface, then for a multilayer with an individual layer thickness of 2 nm (approximately 10 atomic layers), solubility of the composite is effectively on the order of several atomic %.

A summary of experiments performed in our team to document bubble nucleation in Cu-Nb multilayers is presented in table 1. Ion implantation results in a non-uniform concentration profile of the implanted species with a maximum in concentration occurring at a depth that depends on the implantation energy. Note that implanted helium doses of 6 x 10^{16} /cm^2 at 33 keV and 1 x 10^{17} /cm^2 at 150 keV produce equivalent maximum helium concentrations, but at different depths of 120 nm and 450 nm respectively. For these room temperature implantations, the doses we used resulted in two different maximum helium concentrations: (i) 5 at.% in Cu and 7 at.% in Nb, and (ii) 13 at.% in Cu and 21 at.% in Nb. At the 5-7% maximum He concentration, no bubbles were detected in through-focus transmission electron microscopy (TEM) images in the 2.5 nm Cu-Nb multilayers. Also, high-resolution TEM imaging revealed no amorphization or gross changes in interface structure indicating lack of any significant ion-beam mixing between Cu and Nb. However, monolithic films and 100 nm multilayers exhibited 1-2 nm size bubbles. At higher doses (> 13 at.% helium in Cu), bubbles are observed at all layer thicknesses. At the very smallest layer thickness (< 5 nm), helium bubbles were found to decorate the interfaces. In all cases of bubbles nucleated at room temperature, the bubble size was on the order of 1-2 nm.

Retarded Growth of Bubbles

The room temperature implanted samples that had > 13 at.% peak helium concentration and exhibited the 1-2 nm size bubbles were subjected to vacuum annealing at 600 °C for an hour [2]. The monolithic films and the 40 nm multilayers exhibited unstable bubble growth leading to blistering of the films. These blisters were relatively large, on the order of 1-5 μm in diameter, and most were cracked indicating release of helium from the samples. However, in the 5 nm multilayers the bubbles were much smaller. The bubble diameter was 5-10 nm, i.e., on the order of the layer thickness but no unstable growth leading to film blistering was observed. This observation highlights the unusual stability of the 5 nm multilayers subjected to extremely high doses of implanted helium and elevated temperatures.

In a different experiment [5], 33 keV $^4He^+$ ions were implanted at 490 °C to a total dose of 1 x10^{17} /cm^{-2} in two different multilayers: 120 nm and 4 nm. In the 120 nm samples, large faceted bubbles, most exceeding 20 nm in diameter, were observed in the Cu layers. However, in the Nb layers the bubbles remained small, ~1-2 nm in diameter. In the 4 nm samples, the Nb layers still showed the 1-2 nm size bubbles but the size of the bubbles in the Cu layer was limited by the layer thickness of 4 nm. So, there emerges the very interesting behavior that, bubbles in the Cu layer grow to the thickness of the layer and stop at the interface.

In summary, we find that small He bubbles are the first to nucleate. Their sizes remain about 1-2 nm while their number density increases. Eventually the size of the bubbles increases with continued implantation of He, presumably because the He internal pressure in the small bubbles exceeds a critical pressure needed to initiate the punching of interstitials and/or Frank loops. For the conditions of implantation we chose in this study, this happens much more readily in the copper than in the niobium, consistent with the fact that the activation energy for point defect formation is much higher in Nb than in Cu. Thus, at a given temperature, a higher dose of helium is needed for bubble growth in Nb than in Cu.

Table 1 Summary of helium bubble nucleation as a function of room temperature implanted helium dose and layer thickness of multilayers

Irradiation conditions	Peak helium concentration	DPA	Were bubbles observed in TEM?
33 keV, 1.5 x 10^{17} /cm^2 [2]	13 at.% in Cu 21 at.% in Nb	Cu: 35 Nb: 15	5 nm Cu-Nb: yes 40 nm Cu-Nb: yes Single layer Cu: yes Single layer Nb: yes
*33 keV, 6 x 10^{16} /cm^2 or, *150 keV, 1 x 10^{17} /cm^2 [4]	5 at.% in Cu 7 at.% in Nb	Cu: 9 Nb: 6	2.5 nm Cu-Nb: no 100 nm Cu-Nb: yes Single layer Cu: yes

Suppression of Irradiation Hardening

Nanoindentation was used to measure the hardness of ion irradiated Cu-V and Cu-Nb multilayers. The hardness of ion-irradiated films was higher than as-deposited films but only for layer thickness greater than approximately 25 nm. At lower layer thickness the irradiation hardening was significantly decreased and was found to be negligible at layer thickness < 5 nm. For example, in the case of Cu-V, the measured hardness increase after irradiation decreased from ≈ 750 MPa at layer thickness of 200 nm to ≈ 50-100 MPa at layer thickness of 2.5 nm. Radiation hardening in bulk metals caused by radiation-induced defects such as stacking fault tetrahedral, interstitial loops, helium bubbles, etc. In multilayers, as the density of radiation-induced defects and helium bubbles decreases with decreasing layer thickness, radiation hardening diminishes.

Atomistic Modeling

To elucidate the role of interfaces in reducing radiation damage in Cu-Nb multilayer composites, atomic-scale simulations have been carried out using embedded atom method (EAM) potentials [7]. These potentials were constructed by developing a cross-interaction term between well-tested single-element potentials for Cu [8] and Nb [9]. The resulting model of the Cu-Nb binary system correctly reproduces the known dilute enthalpies of mixing of these two elements as well as certain structural and energetic properties determined by first principles simulations using VASP [10]. The hard-core repulsive interactions between the element types modeled by this potential were obtained from the ZBL "universal" potential [11], which has been proven to accurately describe atomic interactions at short distances [12].

Collision cascade simulations were carried out on perfect crystalline FCC Cu and BCC Nb as well as on model Cu-Nb multilayer configurations constructed so as to obey the experimentally observed Kurdjumov-Sachs orientation relation [13]. The layer thicknesses in these configurations were ~4 nm, corresponding to those of the Cu-Nb multilayer composites that were He implanted and studied by TEM. The primary knock-on atom energies used in these simulations ranged from 0.5 to 2.5 keV, well below the approximate threshold energy for ion stopping due to electronic excitation [14]. These cascade simulations revealed a clear difference between the amounts of radiation damage—in the form of vacancy-interstitial (Frenkel) pairs— sustained by perfect crystalline Cu or Nb compared to the Cu-Nb multilayer composite. While all collision cascades occurring in the prefect crystals resulted in the creation of radiation damage, the number of point defects remaining after collisions in the vicinity of Cu-Nb interfaces was always markedly smaller. No point defects remain in the Cu or Nb layers neighboring a Cu-Nb interface as all the vacancies and interstitials created in the cascade core were quickly absorbed into the interface, where they recombined.

The unique ability of Cu-Nb interfaces to trap and recombine Frenkel pairs created during irradiation-induced collision cascades can be traced back to the unusual properties of interfacial point defects in Cu-Nb multilayer composites. One of the most noticeable differences between the behavior of such defects and the ones present in perfect crystalline environments of the corresponding element is the greatly reduced, by almost an order of magnitude, formation energy of interfacial point defects compared to the latter. Thus, vacancies and interstitials that migrate to the Cu-Nb interfaces are effectively trapped there and undergo accelerated recombination due to the enhanced diffusivity and effective size of interfacial point defects. The origin of the unusual properties of interfacial point defects such as low formation energies can be traced back to the structural characteristics of Cu-Nb interfaces, particularly their ability to support a multiplicity of distinct atomic arrangements [10] coexisting in the same interface.

Atomistic modeling is also useful in elucidating the resistance of the interface to the transmission of a single glide dislocation that is the critical unit process that largely determines the maximum strength achieved in nanoscale multilayers. For the case of Cu-Nb, it was shown that interfaces have low shear strength and large in-plane anisotropy of shear strength. This has significant implications for the interactions of glide dislocations, from either Cu or Nb crystal, with the interface. The stress field of a glide dislocation approaching the interface can exert enough stress to locally shear the 'weak' interface, resulting in its absorption and core spreading in the interface plane. Increasing the applied stress in the molecular dynamics simulation does not lead to compaction of the dislocation core and subsequent transmission across the interface. Thus, core spreading effectively pins a glide dislocation in the interface plane and so an incoherent interface that is 'weak' in in-plane shear can be a very strong barrier to slip transmission across interfaces. The weaker the interface, the larger the magnitude of core spreading, and hence, stronger the resistance to slip transmission. Transmission of slip across the interface appears to require multiple glide dislocations that absorb in the interface to react and emit glide dislocations across the interface. In the absence of any stress concentration from a pile-up, such processes may occur at applied stresses on the order of 2.5 - 3 GPa, consistent with the experimentally measured maximum flow strengths of Cu-Nb at h < 5 nm. The results are summarized in Fig. 1.

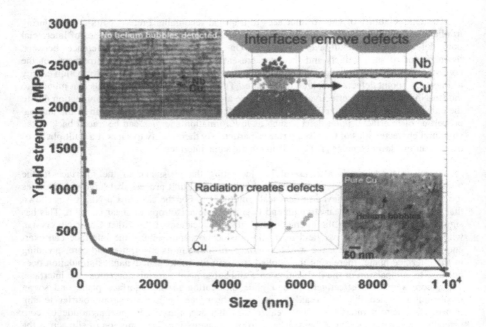

Figure 1: Interfaces act as obstacles to slip and sinks for radiation-induced defects. Nanolayered composites such as Cu-Nb provide orders of magnitude increase in strength and enhanced radiation damage tolerance compared to bulk materials. In the example shown, pure Cu and Cu-Nb (5 nm bilayer period) were irradiated at room temperature with He+ ions to a peak concentration of ≈ 5 at.%. While Cu sample showed defect agglomerates, no helium bubbles were resolved in through-focus imaging in a TEM. Atomistic modeling shows that radiation creates point defects that agglomerate in bulk materials but are attracted, absorbed and annihilated at incoherent interfaces in nanolayered composites. The arrows indicate the corresponding strength levels, prior to irradiation, for the pure Cu and Cu-Nb (5 nm bilayer period) samples.

SUMMARY

Nanolaminate composites can be designed via tailoring of length scales and atomic structure and energetics of interfaces to produce ultra-high strengths and enhanced radiation damage tolerance. This design philosophy was demonstrated using a model Cu-Nb nanolayered system. The flow strengths of these composites are on the order of 2.5 GPa at layer thickness of around 2

172

nm [15] and the best radiation damage tolerance is also observed at the smallest layer thicknesses of a few nanometers. It is well known that the strength increases with decreasing grain size (or, layer thickness) according to the Hall-Petch scaling law. However, the advantages for radiation tolerance are not realized until the relevant length scale, such as the layer thickness, is reduced to the nanometer range. It appears then that the enhanced radiation damage tolerance in nanocomposites is a consequence of short diffusion distance to the nearest sink. At the smallest sizes in the layered composites of a few nanometers, diffusion distances to sinks are short enough, due to the huge interface area in the material, to enable rapid removal of the point defects before they can form into relatively stable aggregates. Of course, the geometry of the nanostructured material must be stable under the extreme irradiation condition. As shown in this work on Cu-Nb nanolayers the layered geometry with flat interfaces extending throughout the sample thickness may provide a benefit in this regard as compared to nanostructured metals with equiaxed grain morphology [16, 17, 18] that may rapidly coarsen under irradiation at elevated temperatures. It should be possible to tailor the atomic structure and energetics of interfaces to provide the most effective sites for point defect trapping and annihilation [10, 19]. Based on the work so far it appears that the key attributes of interface are: (i) multiplicity of atomic arrangements with nearly degenerate energies, (ii) low formation energy of point defects, and (iii) delocalized point defect cores at interfaces such that the recombination distances are large. This is currently being explored via atomistic modeling and experiments in our laboratory. Specifically, the ongoing simulation research aims to create a theoretical framework for predicting the types of interfaces that also possess the structural characteristics observed in the model CuNb system and are therefore good candidates for use as point defect.

ACKNOWLEDGEMENTS

Research at LANL is supported by the Office of Basic Energy Sciences, Department of Energy. Authors acknowledge discussions with J.D. Embury, F. Spaepen and J.P. Hirth.

REFERENCES
1. BES Workshop report *Basic Research Needs for Advanced Nuclear Energy Systems* http://www.science.doe.gov/bes/reports/files/ANES_rpt.pdf
2. T. Hochbauer, A. Misra, K. Hattar and R.G. Hoagland, J. Appl. Phys., **98**, 123516 (2005).
3. M.J. Demkowicz, Y.Q. Wang, R.G. Hoagland and O. Anderoglu, Nucl. Instru. Methods B, **261** (2007), 524.
4. X. Zhang, N. Li, O. Anderoglu, H. Wang, J.G. Swadener, T. Hochbauer, A. Misra and R.G. Hoagland, Nucl. Instru. Methods B, **261**, (2007), 1129.
5. K. Hattar, M.J. Demkowicz, A. Misra, and R.G. Hoagland, Scripta Mater., **58** (2008) 541.
6. A. Misra, H. Kung and R.G. Hoagland, Philos. Mag., **84**, 1021 (2004).
7. M. S. Daw and M. I. Baskes, Physical Review B **29**, 6443 (1984).
8. Y. Mishin, M. J. Mehl, D. A. Papaconstantopoulos, et al., Physical Review B (Condensed Matter and Materials Physics) **63**, 224106 (2001).
9. R. A. Johnson and D. J. Oh, Journal Of Materials Research **4**, (1989) 1195.
10. M. J. Demkowicz, J.P. Hirth and R. G. Hoagland, Phys. Rev. Lett., **100**, (2008) 136102.
11. J. F. Ziegler, J. P. Biersack, and U. Littmark, *The stopping and range of ions in solids* (Pergamon, New York, 1985).

12. K. T. Kuwata, R. I. Erickson, and J. R. Doyle, Nuclear Instruments & Methods in Physics Research Section B-Beam Interactions with Materials and Atoms **201**, (2003) 566.

13. P. M. Anderson, J. F. Bingert, A. Misra, J.P. Hirth, Acta Materialia **51**, (2003) 6059.

14. G. J. Dienes and G. H. Vineyard, *Radiation efects in solids* (Interscience Publishers, New York, 1957).

15. A. Misra, J.P. Hirth and R.G. Hoagland, Acta Materialia, **53**, (2005) 4817.

16. N. Nita, R. Schaeublin, M. Victoria and R.Z. Valiev, Philos. Mag., **85**, (2005) 723.

17. M. Samaras, P.M. Derlet, H. van Swygenhoven and M. Victoria, Phys. Rev. Lett., **88**, (2002) 125505.

18. M. Rose, A.G. Balogh and H. Hahn, Nucl. Instru. Methods B, **127/128**, (1997) 119.

19. H.L. Hienisch, F. Gao and R.J. Kurtz, J. Nucl. Mater., **329-333**, (2004) 924.

Mater. Res. Soc. Symp. Proc. Vol. 1188 © 2009 Materials Research Society   1188-LL08-01

Magnetotactic Bacteria–A Natural Architecture Leading
From Structure to Possible Applications

K. Yu-Zhang[1], K.-L. Zhu[2], T. Xiao[2] and L.-F. Wu[3]

[1] Laboratoire de Microscopies et d'Etude de Nanostructures (LMEN), Université de Reims, France
[2] Key Laboratory of Marine Ecology and Environmental Sciences, Institute of Oceanology, Chinese Academy of Sciences, Qingdao, China
[3] Laboratoire de Chimie Bactérienne, UPR9043, CNRS, Marseille, France

ABSTRACT

Magnetotactic bacteria are aquatic micro-organisms which have the specific capacity to navigate along the lines of the earth's magnetic field. This property is related to the formation of chains of magnetic crystals called *magnetosomes*. All magnetotactic bacteria synthesize nano-sized intracellular magnetosomes that are surrounded by ultra-thin bio-membranes. The magnetosome chains serve as compass for navigation of the magnetotactic bacteria, and the cell flagella are considered as the mechanism for propelling the bacteria forward. This presentation describes various functions of the architectured structure of magnetotactic bacteria as well as their possible applications in biotechnology.

INTRODUCTION

Efficient design in natural systems can be considered in terms of the relationship between form or detailed structure and function just as there is a relationship between the engineer's blueprint and the functioning of a complex structure such as a large building [1]. This approach to the architecture and design of natural systems is reflected in a number of classic texts. A striking example of that is a natural organism (aquatic prokaryote), which builds a machine (a chain of magnetic crystals named *magnetosomes*) directly within its own body. The structure of these tiny bacteria, inhabitants of the first rung on the traditional ladder of life, can illustrate fascinating features within a few microns that some organisms require meters to express.

It was in 1975 that University of New Hampshire microbiologist R. P. Blakemore first discovered this special kind of bacteria in sediments near Woods Hole, Massachusetts (USA) [2], although in a much earlier internal report (1963) the "magnetosensitive bacteria" had been described by S. Bellini at the Institute of Microbiology, University of Pavia (Italy) [3-5]. These bacteria align themselves and migrate in preferred direction along the earth's magnetic field lines, thus the name *magnetotactic bacteria* (MTB). It is well known that such property of MTB is related to the formation of chains of magnetosomes within the cell. Magnetosomes, the single crystalline magnetic particles of 30-120 nm in size, are usually made of ferrimagnetic magnetite (Fe_3O_4, spinel structure, Fd3m space group and lattice parameter a = 0.839 nm), and sometimes though more rarely of greigite (Fe_3S_4, spinel structure with lattice parameter a = 0.988 nm) [6].

If the inorganic magnetosome chains serve as the compass for navigation of magnetotactic bacteria, the organic cell flagella are normally considered as the mechanisms for propelling the bacteria forward. Thus we may ask a number of questions such as how are the mixtures of organic and inorganic systems assembled? What is the basic chemistry of the precipitation of the magnetic particles within a biological environment? And how does the magnetic structure modulate the swimming motion? In order to seek answers to these questions we must examine structures at a range of length scales. And the synthesis of the resultant observations is complex.

Despite the discovery over three decades ago that attracted much attention in the past few years thanks in part to the successful isolation of new strains, and the significant progress in the elucidation of magnetosome biomineralisation, many questions remain to be answered. Here, we try to describe and analyze some functions of the structure of magnetotactic bacteria as well as their possible applications in biotechnology. Thus the approach in the current work is from the viewpoint of the architectured materials by considering the magnetotactic bacteria as a system rather than the more detailed biological viewpoint. Specific review papers on the connection between structure, organisation, and magnetic properties within MTB, as well as on the elucidation of the molecular, biochemical, chemical and genetic bases of magnetosome formation can be found in the references [7-12].

EXPERIMENTAL DETAILS

Sample collections

The MTB samples were taken both from the Mediterranean Sea in the Ile des Embiez, France and from China Sea near the city of Qingdao.

Samples of sediment together with the seawater, with a ratio about 1:2, were first collected from these places and transferred into glass bottles. The bacteria were then enriched by facing the south pole of a permanent magnet (~ 0.4 tesla) outside the bottle 1 cm above the sediment surface. After 30 min, the bacteria attracted by the magnetic field were harvested with a pipette and suspended again into sterile seawater for subsequent analysis.

Transmission electron microscopy observations

Transmission electron microscopy (TEM) is one of the main tools for investigation of the magnetotactic bacteria, particularly for the study of magnetosomes. The MTB cells suspended into sterile seawater can be easily absorbed onto Formvar-carbon coated copper grids for TEM observations directly. Cells were sometimes negatively stained with 1% uranyl acetate for 1 min in order to increase the contrast in the sample in TEM [13].

Routine morphological observations were performed using a Zeiss EM9 microscope operated at 80 kV accelerating voltage. Selected samples were subsequently examined with a TEM (Topcon 002B) equipped with energy dispersive X-ray spectroscopy (EDXS) and high-resolution pole pieces for detailed morphology, crystal structure and chemical composition analyses.

OVERVIEW OF MAGNETOTACTIC BACTERIA AND MAGNETOSOME FORMATION

As mentioned above, magnetotactic bacteria (MTB) are micro-organisms that have the specific capacity of navigation along magnetic field lines. They belong to a heterogonous group of motile prokaryotes, which are ubiquitous in aquatic habitats and cosmopolitan in distribution. A variety of the cell morphologies has been observed, which include rod-, vibrio-, ovoid-, spirillum-, and coccoid forms as well as multicellular bacteria. Figure 1 gives some examples of the MTB morphologies and magnetosome chains observed in the marine MTB from both Ile des Embiez -France and Qingdao -China.

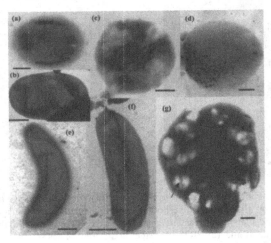

Figure 1. TEM observations of different MTB morphologies and magnetosome chains (a) an ovoid with two chains of rectangular magnetosomes, (b) an ovoid with unusual magnetosome alignment, (c) a coccus with multiple chains of elongated magnetosomes, (d) a coccus with agglomerated magnetosomes, (e) a vibrio with one chain of irregular-shaped magnetosomes, (f) a vibrio with one chain of "shark tooth"-like magnetosomes, and (g) multicellular bacteria with numerous magnetosomes in bullet shape (arrows). Scale bars in the images represent 0.5 μm, except for (g) the bar represents 1.5 μm.

Magnetosomes are the key components of magnetotactic bacteria, as they are dedicated organelles specific for the magnetotactic lifestyle [11]. It is observed that despite the fascinating diversity of both the MTB cellular morphologies and the individual crystal shapes, the magnetosomes have several features in common: 1) they are of high chemical purity - all were identified as the iron oxide magnetite (Fe_3O_4), 2) they are single crystals and almost free of crystalline defects (only twins were sometimes observed), 3) they are single-domain ferrimagnetic particles, 4) their size distribution is relatively narrow, 5) they normally align themselves along the long axis of the cells (with some exceptions). Figure 2 shows some detailed crystalline structure characterisations related to these common features.

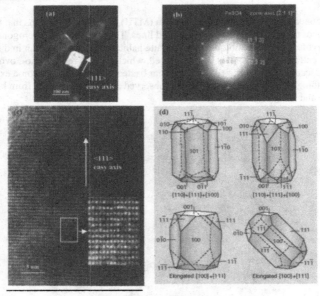

Figure 2. Characterisation of the magnetosome structure. (a) dark-field image along with (b) the corresponding electron diffraction pattern both indicating the nature of the magnetosomes as magnetite (Fe_3O_4) and showing the easy magnetization axis <111> parallel to the elongated direction of the magnetosome, (c) HRTEM images at different scales showing the perfect atomic ranging within individual Fe_3O_4 particle. (d) possible projections of the elongated magnetosome crystals as combinations of {111}, {110} and {100} plans [14].

MAGNETIC ANISOTROPY OF MAGNETOSOMES

Figure 3 shows the size and shape distributions of the magnetosomes and their correlation with the magnetic properties. By analysis of ~100 magnetosomes precipitated within the marine cells at mature stage, we found that the average length and width of the MO1 magnetosomes is about 55 nm and 49 nm, respectively and the majority of the cultured crystals have a slightly elongated cubo-octahedral shape, thus the shape factor (or axial ratio, defined as the ratio of the crystal width/length) is close to 1.0 as shown in figure 3(b) ; on the other hand, the average dimensions of the uncultured QH1 are 100 nm in length and 83 in width, thus the magnetic anisotropy due to the crystalline anisotropy is more obvious in Qingdao's samples.

Theoretical estimates of size and shape ranges of single-domain, two-domain, and superparamagnetic states for magnetite at room temperature were calculated by Butler and Banerjee in 1975 [17] and shown in figure 3(e). This theoretical model makes two main predictions : 1) for an isotropic cube, the critical size for a transition from single-domain to two-domain is about 76 nm, and 2) with increasing crystal anisotropy, the critical size increases and becomes greater than 1000 nm for an axial ratio less than 0.2. Therefore, almost all magnetosomes exhibit single-domain behaviour whatever their shape factor is, as shown in figure 3(e). This means that magnetosomes are indeed uniformly magnetized, which is, *a priori*, required for optimising magnetotaxis (see the next paragraph) in the geomagnetic field. The fact that a number of single-domain magnetite crystals arranged to be in chain and are oriented with the <111> easy axis parallel to the chain axis maximizes the total dipole moment M_T of the cell, thus minimizes the magnetotactic energy. In order to orient the MTB in the geomagnetic field, the magnetic energy of the cell $M_T B_0$ should be larger than the thermal energy $k_B T$, where B_0 is the magnetic induction of the geomagnetic field, k_B the Boltzmann constant, and T the temperature.

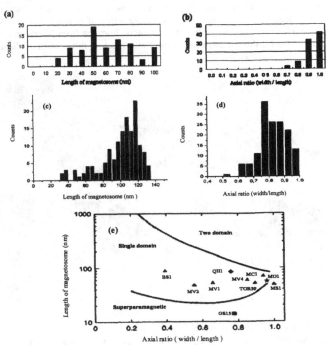

Figure 3. Size and shape distributions of the magnetosomes in the marine MTB (a) size distribution of the cultured MO1 strain (Marseille, France [15]), (b) shape distribution of MO1, (c) size distribution of the uncultured HQ1 strain (Qingdao, China [16]), (d) shape distribution of QH1, (e) Butler-Banerjee diagram of domain states for parallelepiped-shaped magnetite grains (after Bazylinski and Moskowitz [8]). The symbols represent the average sizes of magnetosomes formed by MTB strains such as BS1, MV1, MV2, MV4, MC1, MS1, GS15, TOR39 [18] plus MO1 and QH1 of this work.

Based on this simplified condition, magnetotaxis behaviour needs about 2.1×10^{-13} emu/cell as total dipole moment to counterbalance the thermal energy at room temperature [19]. Our magnetic hysteresis loop measurements of Qingdao's samples give the value of 4.6×10^{-12} emu/cell, which is clearly large enough for magnetotaxis of the bacteria.

MAGNETOTACTIC NAVIGATION APPARATUS

Magnetotaxis refers to the orientation and migration of cells along the magnetic field lines. The question of how the assembly of the magnetosome chain (magnetic compass) and the flagella (propeller) make a navigation apparatus is not yet elucidated. In general, the magnetosome chain aligns along the long axis of rod-shaped or vibroid or spiral cells. If the cell has one unique flagellum at its extremity (Fig 4a), it swims toward only one direction. Hence, the magnetic field determines both the alignment and the direction of its motion (a behaviour known as *polar magnetotaxis)*. It is known that the magnetotactic bacteria from Northern hemisphere swim preferentially toward the geomagnetic North Pole, whereas those from the Southern hemisphere toward the South Pole.

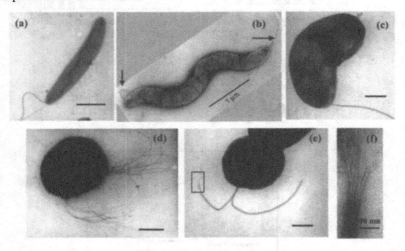

Figure 4. Different configurations of the MTB flagella
(a) a rod-shaped cell with one long flagellum at one cell extremity, (b) a spirillum cell with one long flagellum at each extremity of the cell (arrows, [20]), (c) a vibrio cell with one sheathed flagellar bundle, (d) a coccus with two flagellar bundles [16], (e) a coccus with two sheathed flagellar bundles, (f) the amplified image of the rectangular part of (e) showing details of flagella within a sheath. The scale bars represent 0.5 μm except for (b) and (f) [15].

On the other hand, if MTB possess one flagellum at each cell extremity (Fig 4b), they can swim toward both directions. In this case, the magnetic field determines only the magnetosome alignment but no longer the motion direction (a behaviour known as *axial magnetotaxis*). Although magnetotactic cocci exhibit polar magnetotaxis behaviour, their navigation apparatus may follow a different mechanism compared to rod-shaped or vibroid bacteria. In fact, the magnetotactic cocci usually have two bundles of flagella located on the same side of the cells (Fig. 4d and e). Their magnetosome chains can be either parallel or perpendicular to the line linking the bases of the two flagellar bundles. Normally, magneto-cocci rotate on themselves (like a spinning top) while swimming. Since the axes of cell translation, rotation and magnetosome chain are not over-lapped, the cocci exhibit a helical swimming trajectory [15].

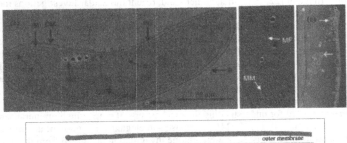

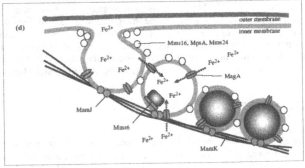

Figure 5. Architecture of magnetotactic bacteria
(a) cryo- electron tomography (CET) revealing that magnetosomes are invaginations of the inner membrane (OM: outer membrane; IM: inner membrane; PG: peptidoglycan layer; CR: chemoreceptor chain; MG: magnetosome chain; PHB: poly-β-hydroxybutyrate granule; R: ribosimes; B: outer membrane bleb; G: gold fiduciary marker [22]), (b) organisation of the magnetosome chain and the cytoskeletal magnetosome filament (MF) shown by CET [23], (c) tomographic reconstruction showing the cytoplamic membrane (blue), empty magnetosome membranes (MM, yellow), growing and mature magnetite crystals (red), and the MF (green) [23], (d) Schematic of possible mechanism leading to magnetite biomineralisation and the roles of the proteins that were so far shown to be necessary for magnetite biomineralisation [12].

POSSIBLE MAGNETITE BIOMINERALISATION MECHANISMS

As early as in 1980, D. Balkwill and R. P. Blakemore have already showed by conventional TEM the evidence of the magnetosomes individually surrounded by a lipid bilayer named late as *magnetosome membranes* [21]. However, significant progress in observations of the detailed cell structure was truly made in 2006 owing to the development of cryo-electron tomography (CET) technique. As shown in figure 5, CET studies revealed not only various components of a MTB cell but also one important fact that both empty magnetosome membranes and the vesicles that contain growing magnetosomes are closely attached to the cytoskeletal magnetosome filament [22, 23]. On the other hand, recent molecular investigations including genome sequence, gene expression, mutagenesis, and proteome analyses successfully identified a number of genes and proteins which helped the magnetosome formation within MTB cell. For example, MamK is known as a filamentous protein that supports the chain–like structure, MamJ is also a filamentous structure which directs the assembly and maintenance of the chain; the MagA protein seems to play a role in iron efflux in the cytoplasmic membrane and in iron influx in the magnetosome membrane, while the Mms16, MpsA and Mms24 all help in the activation of the vesicle formation; finally the Mms6 protein is important in control of the shape and size of magnetite crystal formation, as shown in figure 5 (d) [12]. Based on the above results, several models were suggested for understanding the mechanisms of magnetite biomineralisation within magnetotactic bacteria [10, 11, 12, 24]. A concise and elegant description of this process was given by Komeli in three steps: first, a membrane invagination is derived from the inner membrane, and magnetosome proteins are sorted away from cell membrane proteins. Second, individual invaginations are assembled into a chain with the help of MamJ and MamK proteins. Third, iron is transformed into highly ordered magnetic crystal within the magnetosome membrane with the possible involvement of genes from the Mam CD and Mms6 operons [24].

POTENTIONAL APPLICATIONS

The magnetic crystal formation in the magnetotactic bacteria has the potential to yield useful biomaterials. Compared with inorganic synthesis techniques, biomineralisation process can provide a way to produce highly uniform and nano-scaled crystals [25]. Their perfect crystalline structure, narrow size distribution and specific magnetic properties make the magnetosomes potentially useful in a number of applications, such as for magnetic drug targeting, magnetic separation, diagnostics, or as a contrast agent for magnetic resonance imaging [26, 27]. And the fact that each magnetosome is enveloped by a biological membrane (MM) can prevent from their agglomeration. There is no coarsening reaction as would be expected from free arrays of small crystallites.

The unique and superior characteristics of magnetosomes seem to be in their use as magnetic nanoparticles [11]. However, their real applications also rely on the availability of large amounts of bacteria at reasonable costs. In order to be truly competitive with the artificial fabricated magnetite particles [28], there is still much to do perhaps by exploring more the use of whole magnetotactic bacteria. Some suggested applications of these involve the use of living and actively swimming cells for the removal of heavy metals and radio-nuclides from waste water [29, 30]. This is quite an attractive and challenging idea.

Due to their magnetic alignment in the geomagnetic field, magnetosome chains potentially contribute after death to the natural magnetisation of the sediment which then become magneto-fossils [31]. Further research will be required to understand the stability of magnetosomes in the sediment in order to use the magneto-fossils as reliable paleo-tracers [32]. Finally, magnetite crystals with morphologies and sizes similar to those in MTB have been reported in the Martian meteorite ALH84001, which raised the possibility that nanoscaled magnetite could serve as a biomarker for life beyond the earth [33].

DISCUSSION

In a broad sense, we can consider the development of the magnetotactic structures in bacteria as a specific example of biomineralisation. The topic of biomineralisation has been reviewed in detail in two classical texts [34, 35]. The process of biomineralisation is characterized by a number of features including rapid reaction kinetics at room temperature and often the precipitation of a mineral phase in conjunction with the action of proteins which control both the nucleation morphology and growth kinetics of the mineral phase. Chemical precipitation of highly crystalline particulate magnetite is usually carried out at high temperatures, with ambient temperature reactions being slower and generally producing poor quality crystals [36]. In contrast, it is well known that biological systems catalyze reactions at physiological temperatures that would require high temperatures for purely chemical conversion to overcome large activation barriers. Staniland *et al.* showed by real-time X-ray magnetic circular dichroism that magnetite crystals appear to be of a mature size and well defined morphology within a cell, taken just 15 min after induction [37]. Hence, the biomineralisation of magnetite occurs rapidly at room temperature in MTB on a similar time scale to high-temperature chemical precipitation reactions. It is clear that the magnetite precipitation in MTB is caused by a biological catalysis of the process.

As shown in figure 5(d), various proteins associated with the magnetosome membranes could play functional roles involved in the magnetite precipitation, such as the accumulation of supersaturating iron via the MagA, the heterogeneous nucleation of the iron oxides activated by the Mms6 and supporting the chain-like structure with MamK and MamJ. It is also worth emphasising the role played by the magnetosome membrane in maintaining the magnetosome chain formation. It is observed that isolated magnetite crystals were tightly interconnected by the organic material even when they were separated far enough to have no significant dipolar attraction effect [11]. This means that the formation of chains requires the presence of the magnetosome membrane which provides spacing and seems to mediate contact between adjacent magnetosomes [23]. Elucidation of the precise roles of the genetic and biochemical determinants in biomineralisation and in molecular mechanisms of magnetosome formation is one of the future issues.

The functionalisation is important because it links the inorganic entity, *i.e.* the magnetosome crystal, to the living structure in terms both of bonding to the structure but also how it can go from receptor to control of the motion of the flagella. In this paper, we focused on the magnetite (Fe_3O_4) formation within magnetotactic bacteria because the majority of magnetosomes are composed of Fe_3O_4. However, greigite (Fe_3S_4) magnetosomes also exist. If magnetite-type MTB prefer to live in microaerobic to anaerobic conditions, the living of greigite-type MTB require the presence of sulphide in anaerobic conditions. Therefore, magnetite producers occur higher in the water column than greigite producers [38]. It is amazing to see that both magnetite- and greigite-type magnetosomes can co-precipitate within a same cell [39], which makes the biomineralisation process even more complicated to understand. The mineral and chemical compositions of magnetosomes appear to be strictly controlled by the organism, which often favours magnetite formation whatever the incorporated elements found within MTB. For example, we detected by EDXS technique phosphate-like granules within the samples from Mediterranean Sea [15] and sulphur-rich granules within those from China Sea (dark-contrast granules in figure 1(a) [16]); the others reported the presence of gold, silver, calcium and barium [40, 41], all within magnetite-producing MTB. Moreover, all successfully cultured strains can only synthesize magnetite even when incubated under reducing conditions that should favour the formation of greigite [42]. Therefore, the pathway of magnetosome formation seems to have a very high specificity for iron.

Although the roles of all magnetosome-associated proteins and the exact biomineralisation process have not been fully identified, it is clear that magnetosome formation is highly regulated by the organism with regard to iron transport, nucleation, crystal growth, morphology, and chain formation. As a highly regulated process that has evolved to produce minerals with specific structures and functions, it is generally characterised by 1) uniform particle sizes, 2) complex morphologies, 3) well-defined structures and organisation, and 4) higher order assembly into hierarchical structures [42].

Magnetotactic bacteria are the simplest single-cell micro-organisms in which biomineralisation occurs. Given the complexity of the biomineralisation process shown in figure 5(d) and discussed above, the magnetotactic bacteria provide an ideal model system to study the biomineralisation mechanisms as well as a good example of naturally architectured multifunctional materials.

SUMMARY

The topic of magnetotactic bacteria has been reviewed elsewhere in a series of articles. What is presented here is a viewpoint of a materials scientist looking at the relationship between structure and functions and examining MTB as an architectured system.

The magnetite size, crystallographic orientation and chain assembly of magnetosomes in MTB are all highly significant for the function of magnetotaxis in the geomagnetic field [8]. In fact, the magnetosome chain is a masterpiece of permanent magnet design that makes each cell, in effect, a self-propelled magnetic dipole. As a compass needle, the chain fits into micron-diametered cell *in situ* and is auto-assembled. The MTB flagella play the role for propelling the bacteria forward by using aerotaxis to efficiently locate and remain at an optimal oxygen concentration area. Despite the fact that there remain a number of fundamental challenges on elucidating the mechanism of magnetosome synthesis, magnetotactic bacteria truly provide a good example both for the investigation of biomineralisation and for the study of architectured materials.

ACKNOWLEDGMENTS

We would like to thank Alain Bernadac, Christopher Lefèvre, Claire Santini, Hongmiao Pan for close collaboration, and Hongqing Wu for paper editing assistance. The authors also gratefully acknowledge the supports from both the Chinese Academy of Sciences (the grant for outstanding oversea chinese scholars 2006-1-15) and the French Centre National de la Recherche Scientifique (Fellowship to Kailing Zhu).

REFERENCES

1. D'Arcy Thompson, *On Growth and Form*, ed. J. T. Bonner, Cambridge Univ. Press (1992).
2. R. P. Blakemore, *Science* **190**, 377 (1975).
3. S. Bellini, http://www.calpoly.edu/~rfrankel/Sbellini2.pdf.
4. S. Bellini, *Chinese J. Oceanology and Limnology* **27**, 3-5 & 6-12 (2009).
5. R. B. Frankel, *Chinese J. Oceanology and Limnology* **27**, 1-2 (2009).
6. M. Pósfai, P. R. Buseck, D. A. Bazylinski, and R. B. Frankel, *Science* **280**, 880 (1998).
7. R. P. Blakemore, *Annu. Rev. Microbiol.* **36**, 217-238 (1982).
8. D. A. Bazylinski and B. M. Moskowwitz, *Microbial biomineralisation of magnetic iron minerals*, ed. J. F. Banfield and K. H. Nealson (1997), *Rev. Mineral.* **35**, pp. 191-223.
9. R. B. Frankel and B. M. Moskowwitz, "Biogenic Magnets", *Magnetism: Molecules to Materials IV*, ed. J. S. Miller and M. Drillon (Wiley-Vch, 2003), pp. 205-231.
10. D. A. Bazylinski and R. B. Frankel, *Nature Reviews* **2**, 217-230 (2004).
11. D. Faivre and D. Schüler, *Chem. Rev.* **108**, 4875-4898 (2008).
12. A. Arakaki, H. Nakazawa, M. Nemoto, T. Mori, and T. Matsunaga, *J. R. Soc. Interface* **5**, 977-999 (2008).
13. C. Lefèvre, A. Bernadac, N. Pradel, L.-F. Wu, K. Yu-Zhang, T. Xiao, J. P. Yonnet, A. Lebouc, T. Song, and Y. Fukumori, *J. Ocean University of China* **6**(4), 355-359 (2007).
14. B. Devouard, M. Pósfai, X. Hua, D. A. Bazylinski , R. B. Frankel, and P. R. Buseck, *American Mineralogist* **83**, 1387–1398 (1998).
15. C. T. Lefèvre, A. Bernadac, K. Yu-Zhang, N. Pradel, and L.-F. Wu, *Environmental Microbiology,* in press (2009).
16. H.-M. Pan, K.-L. Zhu, T. Song, K. Yu-Zhang, C. Lefèvre, S. Xing, M. Liu, S. Zhao, T. Xiao, and L.-F. Wu, *Environmental Microbiology* **10**(5), 1158-1164 (2008).
17. R. F. Butler and S. K. Banerjee, *J. Geophysical Research* **80**, 4049–4058 (1975).
18. Zhang C.-L., H. Vali, C. S. Romanek, T. J. Phelps, and S. V. Liu, *American Mineralogist* **83**, 1409-1418 (1998).
19. R. B. Frankel, *Annu. Rev. Biophys. Bioengng* **13**, 85 (1984).
20. R. B. Frankel and D. A. Bazylinski, *Trends in Microbiology* **14**, 329-331(2006).
21. D. Balkwill and R. P. Blakemore, *J. Bacteriol.* **141**, 1399-1408 (1980).
22. A. Komeili, Z. Li, D. K. Newman, and G. J. Jensen, *Science* **311**, 242-245 (2006).
23. A. Scheffel, M. Gruska, D. Faivre, A. Linaroudis, J. M. Plitzko, and D. Schüler, *Nature,* **440**, 110-114 (2006).
24. A. Komeili, *Annu. Rev. Biochem.* **76**, 351-366 (2007).
25. M. Sarikaya, *Proc. Natl. Acad. Sci. USA* **96**, 14183–14185 (1999).
26. E. J. Baeuerlein, D. Schüler, R. Reszka, and S. Päuser, *In PCT/DE* 98/00668 (1998).

27. T. Matsunaga, T. Suzuki, M. Tanaka, and A. Arakaki, *Trends Biotechnol.* **28**, 182 (2007).

28. M. Ma, Y. Zhang, W. Yu, H.-Y. Shen, H.-Q. Zhang, and N. Gu, *Colloids and Surfaces A* **212**, 219-226 (2002).

29. A. S. Bahaj, I. W. Croudace, and P. A. B. James, *IEEE Transactions on Magnetics* **30**, 4707-4709 (1994).

30. A. S. Bahaj, I. W. Croudace, P. A. B. James, F. D. Moeschler, and P. E. Warwick, *J. Inorg. Biochem.* **59**, 107 (1998).

31. S.-B. R. Chang and J. L. Kirschvink, *Ann. Rev. Earth Planet. Sci.* **17**, 169 (1989).

32. N. Petersen, T. Von Dobeneck, and H. Vali, *Nature* **320**, 611 (1986).

33. K. L. Thomas-Keprta, S. J. Chemett, D. A. Bazylinski, J. L. Kirschrink, D. S. McKay, H. Vali, E. K. Gibson, and Jr. C. S. Romanek, *Proc. Natl. Acad. Sci. USA* **98**, 2164-2169 (2001).

34. S. Mann, *Biomineralisation: Principles and Concepts in Bioinorganic Materials Chemistry*, Oxford University Press (2001).

35. D. Fortin, S. Glasauer, and R.S. Langley, *Biomineralisation: From Nature to Application*, vol. 4 of Metal Ions in Life Sciences, ed. A. Sigel, H. Sigel, R. K. O. Sigel (Jihn Wiley & Sons), in press.

36. R. M. Cornell and U. Schertmann, *The Iron Oxides: Structure, Properties, Reactions, Occurrence and Uses,* VCH, New York (1996).

37. S. Staniland, B. Ward, A. Harrison, G. van der Laan, and N. Telling, *Proc. Natl. Acad. Sci.* **104** (49), 19524-19528 (2007).

38. D. A. Bazylinski, R. B. Frankel, B. R. Heywood, S. Mann, J. W. King, P. L. Donaghay, and A. K. Hanson, *Appl. Environ. Microbiol.* **61**, 3232 (1995).

39. D. A. Bazylinski, B. R. Heywood, S. Mann, and R. B. Frankel, *Nature* **366**, 218 (1993).

40. C. N. Keim and M. Farina, *Geomicrobiol. J.* **22**, 55 (2005).

41. A. Isambert, N. Menguy, E. Larquet, F. Guyot, and J.-P. Valet, *Am. Mineral.* **92**, 621 (2007).

42. F. C. Meldrum, S. Mann, B. R. Heywood, R. B. Frankel, and D. A. Bazylinski, *Proc. R. Soc. Lond. B* **251**, 231-236 & 237-242 (1993).

Mater. Res. Soc. Symp. Proc. Vol. 1188 © 2009 Materials Research Society   1188-LL06-02

Next Generation Phase Change Materials as Multifunctional Watery Suspension for Heat Transport and Heat Storage

M. Hadjieva[1], M. Bozukov[1], I. Gutzov[2]
Bulgarian Academy of Sciences:
[1]Central Laboratory of Solar Energy and New Energy Sources, 72 Tzarigradsko Schosse Blvd., 1784 Sofia, Bulgaria
[2]Institute of Physical Chemistry, Acad G. Bonchev, bl. 11, 1113 Sofia, Bulgaria

ABSTRACT

The phase change materials (PCM) were modified to paraffin watery dispersions, well structured as multifunctional fluids for both, heat transport and heat storage in thermal cooling technology. Two types PCM dispersions, paraffin microcapsule watery suspension (Slurry A) and paraffin microemulsion (Slurry B), absorbed and released heat in a temperature range of phase transition from 2^0C to 12^0C. Thermal capacity of paraffin Slurry A was about 53 kJ/kg, while paraffin Slurry B exchanged about 83 kJ/kg amount of heat during phase transition.

Systematization of thermophysical data and structural imaging of Slurry A and Slurry B, studied before and after thermal cycling, allowed comparison of thermal efficiency and structural stability during multiple phase transitions. The correlation between structural properties and thermal storage capacity allows recommendation of a paraffin emulsion Slurry (B) suitable for practice as a watery dispersion architectured stable in limits of year and thermally efficient.

INTRODUCTION

The PCM, well known since past century, are modified to the next generation of PCM intended to respond to new functions and requirements of the present-day thermal technology. Impact of phase change technology on energy balance of dwellings includes reduced energy consumption of heating/cooling, moderate temperature swings, reduced of peak power needs.

Modern cooling technology developments use the PCM slurry for storage and transport of heat.[1],[2],[3],[4]. PCM slurries are multifunctional fluids composed of microstructures dispersed in a solvent as: phase change watery suspension with micro-capsulated paraffin [3], [4], phase change paraffin emulsion slurry [2], clathrate hydrate slurry.

Advantages of architectured PCM watery dispersions comprise multifuncionality of heat storage and heat transfer processes; tailoring the temperature range of heat storage; improvement of energy storage capacity of heat transfer fluids; structural stability of PCM during thermal cycling at repeatable phase transition; improved thermal conductivity.

PURPOSE

Long-term effectiveness of the architectured PCM watery dispersions, PCM slurries depends on thermal energy storage capacity, on structural stability of PCM at multiple melting/solidification and on structural stability of the PCM slurry during repeatable thermal cycling at heating and cooling. Long-term pumpability and a heat flow transfer of PCM slurry, which proceeds as a heat transfer fluid are of practical importance. Progress in solving the material requirements is crucial for PCM slurry technology developments.

Thus, structural and thermophysical properties of two types modified paraffin dispersions were in the focus of experiments for selection of advantageous multifunctional slurry with structural, kinetic and thermodynamic stability, appropriate for efficient thermal technology.

EXPERIMENTAL

Paraffin slurries are intended to avoid some PCM drawbacks like PCM structural instability or PCM low thermal conductivity and to work as paraffin multifunctional fluids. Our study was concentrated on PCM, modified to the next generation PCM presented by two types PCM watery dispersions: Slurry A - the paraffin watery suspension, prepared of microencapsulated paraffin dispersed in water and Slurry B- the paraffin micro-emulsion, formulated by paraffin oil droplets dispersed in water. Slurry elaboration methods differ from PCM watery suspension to PCM emulsion and cause diversity in the PCM slurry properties.

PCM slurries were studied in a temperature range of phase transition from -5^0C to 20^0C. The differential scanning calorimeter (DSC), NETZSCH; scanning electron microscopy (SEM), JOEL model JSM-5510 at 6,57.10-7Pa vacuum; hot stage optical microscopy, LINKAM for study of phase transition phenomena "in situ" during thermal cycling; a hand made thermal cycling system, where Danfoss cooling machine ensured 2 kW cooling capacity at 40^0C and PCM slurry (2L) circulated with optimum flow rate of 0.1 L/min through hot and cold zones – all equipments produced accurate data for complete characterization from structural and thermal point of view both types of paraffin slurries.

The mechanical, kinetic and thermodynamic potentials of paraffin watery dispersion systems were evaluated in respect to slurry stability for practical technology applications.

PARAFFIN SLURRY STABILITY

Paraffin micro-emulsion stability

Slurry B consists of paraffin oil (32 wt%) dispersed in water. The formulated PCM emulsion was stabilized by surfactants, which prevent separation of both immiscible liquids.

Mechanical stability of Slurry B, pumpability of Slurry B, was studied as a result of multiple circulations in a cycling system at a room temperature. Slurry B circulated smoothly through pumps. Aggregation of paraffin microcapsules or of a paraffin leakage was not observed. Kinetic stability of Slurry B at breakdown processes like creaming, sedimentation, flocculation or coalescence was evaluated from droplet size distribution, defined by density difference between paraffin droplets and water medium. Paraffin droplet size distribution was scanned with a particle size analyzer at X1000 magnification seen on figure 1. The majority of droplets had diameters around and below 1 μm. It indicates that creaming in the Slurry B did not cause coalescence of paraffin oil droplets. Thermodynamic stability was tested by thermal cycling system with 2 litra Slurry B in a temperature range of −10 to 25 ^{0}C and a flow rate of 0.1 L/min. It was proved, that Slurry B remained in dispersion after 500 thermal cycles through of hot and cold zone. Additionally, a thermal microscopy study with a LINKAM hot stage at a heating/cooling rate of 1.5 K/min (table I) showed regular distribution of the paraffin droplets in

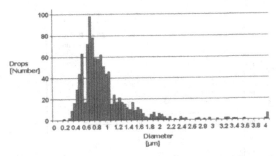

Figure 1. Histogram of the droplet size distribution of Slurry B, tetradecane in water emulsion.

water medium (see table I). Crystallization seems to begin around 3^0C in a droplet phase (table I, a). Below zero additional crystallization was observed between droplets (table I, b) and some water crystals. Crystallization scans in polarized light by crossed nicols (table I, c) let further evaluation of a paraffin particle size (around 1 μm) and homogeneous droplet distribution.

Table I. Thermal optical microscopy of droplet distribution in Slurry B during crystallization.

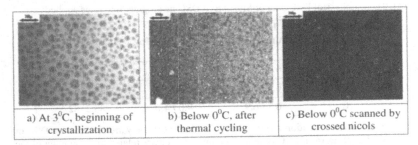

a) At 3^0C, beginning of crystallization	b) Below 0^0C, after thermal cycling	c) Below 0^0C scanned by crossed nicols

A few changes in paraffin spherical drop shape, size and distribution, the single effect of coalescence during melting at high temperatures and single clusters in PCM slurry caused by the surfactant inefficient role in time and temperature, were observed finally after multiple thermal cycling (55 cycles) in a micro volume of 25ml. of a LINKAM hot stage.

Tests on thermodynamic stability of a Slurry B were completed by comparison of phase transition temperatures obtained by a LINKAM hot stage to the DSC data, which (slurry samples volume up to 20ml.) were collected in temperature range of –5 to 20 ^{0}C at a heating/cooling rate of 1.5 K/min. The melting DSC curves were well repeatable, while the small peak appeared at about -2°C on cooling curves. The second small peak at a DSC cooling curve indicates evidently the supercooling effect. Furthermore the DSC measurements show the clear dependence of the thermophysical characteristics on heating/cooling rate applied (see figure 2.).

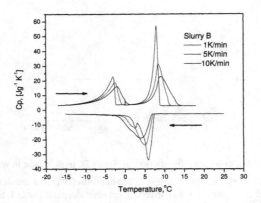

Figure 2. Dependence of the Slurry B thermophysical data on DSC heating/cooling rate applied.

The optimal heating/cooling rate for DSC investigation was selected to be from 1.5K/min to 2.5 K/min, based on analysis of DSC data summarized.(see table II).

Table II. Thermophysical DSC data of Slurry B at DSC heating/cooling rates applied.

Heating/ Cooling Rate [K/min]	Tonset Heating/Cooling [oC]	Tpeak Heating/Cooling [oC]	ΔH Heating/Cooling [kJ/kg]
1	6.1 / 7.4	7.9 / 5.9	84.2/ 83.4
5	6.3 / 7.1	8.5 / 5.0	84.8/ 83.1
10	6.7 / 6.9	9.0 / 4.0	83.2/ 82.5

Temperatures of phase transition fixed by LINKAM hot stage study are closer to the DSC onset temperatures, but differ from DSC peak temperatures, which are related to the amount of the DSC samples. Difference between the amount of heat absorbed and released by Slurry B reached 5 kJ/kg after multiple DSC thermal cycling. DSC data would be used for modeling and simulation and further improvement of PCM slurry in search of better material options.

Paraffin watery suspension stability

Slurry A is elaborated of paraffin microcapsules, 46.5 wt%, dispersed in water as a PCM watery suspension. The mechanical stability or pumpability of Slurry A, experimented during circulation at room temperature by means of most simple centrifugal pumps was satisfactory. Microcapsules remained stable in form and size. Sedimentation or aggregations were not observed. Some sedimentation effects or clusters were found with SEM imaging when Slurry A was left at rest and quantity of water medium was not sufficient for proper suspension (table III, b, c). Kinetic stability of Slurry A was achieved by well-balanced density difference of paraffin microcapsules and water medium. Evidently, the polymeric layer formed the microcapsules ensured good structural stability. The microcapsule size of Slurry A remained from 1.2 to 2.9μm, while a regular microcapsule distribution into medium was observed before/after cycling.

Table III. SEM imaging of the Slurry A mechanical stability after pumping tests.

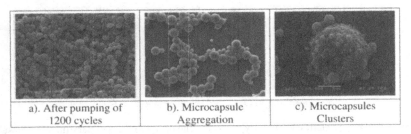

| a). After pumping of 1200 cycles | b). Microcapsule Aggregation | c). Microcapsules Clusters |

Thermodynamic stability was studied by thermal cycling of 2 litra slurry through hot and cold zones, complemented by DSC thermal measurements at 1.5 heating/cooling rate of up to 20 mg slurry. On figure 3 is presented a DSC cycle of Slurry A at melting and solidification process.

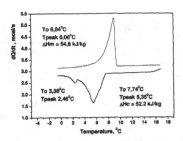

Figure 3. The DSC thermal cycle of paraffin microcapsule watery suspension, Slurry A.

Thermophysical data calculated by NETZCH software are compared on table IV. The DSC hysteresis between onset temperatures of melting/solidification is 1.23 degrees what is acceptable for a cooling practice, although this value would change with larger amount of slurry.

Table IV. Thermodynamic DSC characteristics of paraffin microcapsule Slurry A

DSC Heating/ Cooling Rate [K/min]	Tonset Heating/ Cooling [°C]	Tmax Heating/ Cooling [°C]	ΔH Heating/Cooling [kJ/kg]
1.5	6.83/7.74	8.06/5.35	54.8/52.2

Thermodynamic stability data are important economical characteristics for real time applications so, thermal cycling stability test was done for three months circulation of the Slurry A passing through hot and cold zone of thermal cycling system. The DSC data for heat absorbed /released by Slurry A were 20% lower than DSC data of first DSC cycles what indicates that melting/solidification processes are not sufficiently controlled. The SEM imaging of Slurry A showed regular capsules together with some deformed or broken paraffin microcapsules (table V).

Table V. SEM imaging of Slurry A after thermal cycling of 1500 cycles.

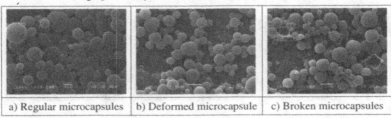

| a) Regular microcapsules | b) Deformed microcapsule | c) Broken microcapsules |

Our investigations clarified the Slurry A success and open problems of paraffin microcapsule stability to be solved prior to development of modern technology in response to expectation of practice.

CONCLUSIONS

The amount of heat from 53 kJ/kg to 83 kJ/kg, absorbed or released during the phase transition of paraffin Slurry A and paraffin Slurry B, melted/solidified in the temperature range from 2^0C to 12^0C degrees, are appropriate for cooling technology development.

Structural images of the paraffin slurries scanned before and after thermal cycling of heating and cooling correspond to the thermophysical behavior. The paraffin micro-emulsion structure is associated with higher thermal storage capacity and better structural fluidity, than experimental data of paraffin microcapsule slurry probably due to the heat transfer process through larger contact surface between dispersed and continuous phase of the emulsion.

Phase change micro-emulsion slurry is a promising, rational way of structuring of PCM fluids for thermal energy storage and transportation technology, when spontaneous reduction of an interfacial area by coagulation or coalescence of PCM droplets is controlled.

ACKNOWLEDGMENTS

Thankfulness to European Community for the EU grant (ENK6-CT-2005-00507), allowed our successful research work and achievements, also thankfulness to all project partners for our wonderful teamwork.

REFERENCES

1. H. Inaba., "New challenge in advanced thermal energy transportation using functionally thermal fluids", *Int. J. Therm. Sci.* **39,** 991-1003 (2000).
2. L. Royon, G.Guiffant, "Heat transfer in paraffin oil/water emulsion involving supercooling phenomenon", *Energy Conv. and Manag.*, **42,** 2155-2161(2001).
3. X. Wang, J. Niu, Y. Li, Y. Zhang, "Microencapsulated phase change material slurry as heat transfer media in a circular pipe", *Proc. Int. Congress of Refrig*, Beijing, China, (2007) ICR07-B1-248.
4. L. Royon, F. Trinquet, O. Bros, P. Mercier, G. Guiffant, "Flow and heat transfer investigations on paraffin stabilized material slurry for district cooling and air conditioning process", *Proc. Int. Congr. of Refr.*, Beijing, China,(2007) ICR07-B1-1484.

Mater. Res. Soc. Symp. Proc. Vol. 1188 © 2009 Materials Research Society

Multifunctional One-Dimensional Phononic Crystal Structures Exploiting Interfacial Acoustic Waves

Albert C. To[1,2] and Bong Jae Lee[1]

[1]Department of Mechanical Engineering and Materials Science, University of Pittsburgh, Pittsburgh, PA 15261

[2]Department of Civil and Environmental Engineering, University of Pittsburgh, Pittsburgh, PA 15261

ABSTRACT

The present study demonstrates that interfacial acoustic waves can be excited at the interface between two phononic crystals. The interfacial wave existing between two phononic crystals is the counterpart of the surface electromagnetic wave existing between two photonic crystals. While past works on phononic crystals exploit the unique bandgap phenomenon in periodic structures, the present work employs the Bloch wave in the stop band to excite interfacial waves that propagate along the interface and decay away from the interface. As a result, the proposed structure can be used as a wave filter as well as a thermal barrier. In wave filter design, for instance, the incident mechanical wave energy can be guided by the interfacial wave to the lateral direction; thus, its propagation into the depth is inhibited. Similarly, in thermal barrier design, incident phonons can be coupled with the interfacial acoustic wave, and the heat will be localized and eventually dissipated at the interface between two phononic crystals. Consequently, the thermal conductivity in the direction normal to the layers can be greatly reduced. The advantage of using two phononic crystals is that the interfacial wave can be excited even at normal incidence, which is critical in many engineering applications. Since the proposed concept is based on a one-dimensional periodic structure, the analysis, design, and fabrication are relatively simple compared to other higher dimensional material designs.

INTRODUCTION

Recently, there has been much interest in material design that can achieve acoustic cloaking [1-3], which makes an object 'invisible' to acoustic waves, and hence has some important civilian and military applications. For example, the phenomenon can be employed to hide a submarine from active sonar detection. Metamaterials having strong mass anisotropy have been suggested for acoustic cloaking [1-3]. Besides cloaking, perfect absorption of acoustic waves without reflection by a material can be employed to hide an object from active sonar detection. The present study demonstrates perfect absorption of acoustic waves by stacking two phononic crystals (PC) together in series at certain frequencies. The absorption is achieved by conversion of the incident bulk acoustic waves into interfacial waves between the two phononic crystals. The interfacial wave existing between two phononic crystals is the counterpart of the surface electromagnetic wave existing between two photonic crystals [4]. While past works on phononic crystals exploit the unique bandgap phenomenon in periodic structures [5,6], the present work employs the Bloch wave in the stop band to excite interfacial waves that propagate along the interface and decay away from the interface.

A peculiar behavior of periodic structures is the existence of the so-called bandgaps or stop bands occurring over specified frequency ranges. Within a phononic bandgap, the propagation of elastic or thermal waves is inhibited if one considers phononic bandgap materials. Phononic bandgaps arise due to the periodic arrangement of one-dimensional (1D) constituent layers with different shear and Young's modules; hence the structure is called a phononic crystal. The bandgap of a phononic crystal can be calculated by using transfer matrix formulation [7-9]. The transfer matrix formulation will result in an angular frequency versus the parallel component of the wavevector plot called the band structure, consisting of pass bands and stop bands, as shown in Fig. 1. In the pass band, for instance, elastic waves can propagate through a phononic crystal, whereas in the stop band, no energy-carrier waves can exist inside a phononic crystal, and only an oscillating but evanescently decaying field may exist. The existence of stop bands makes phononic crystals extremely appealing for engineering applications, such as thermal barriers and mechanical filters.

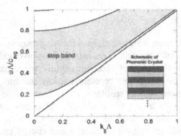

Fig. 1 A typical band structure plot of a phonon crystal consisting of a stop band (yellow color).

Recently, it has been shown that surface plasmon polariton can be generated in the stop bands of photonic crystals [10,11], and its counterpart in phononic crystals is demonstrated in this study. A surface plasmon polariton is an electromagnetic wave that propagates along the interface and decays exponentially from the interface into both media [12,13]. At the resonance condition, the incident photon energy is transferred to the polariton, resulting in a sharp dip in the reflectance spectroscopy. Since surface polaritons involve two evanescent waves at the interface, a prism coupler is often required to excite the surface plasmon. However, it has been shown that an effective evanescent wave (i.e., decaying Bloch wave) also exist in the stop bands of photonic crystals [14-16]. Therefore, a photonic crystal may be used as a polariton coupler so that a propagating wave in air can directly excite the surface polariton. In this work, we demonstrate that the analogous interfacial acoustic wave can be generated at the interface between two different phononic crystals that are bounded together.

ANALYSIS

Figure 2 shows a 1D phononic crystal (PC) structure consisted of two phononic crystals (PC) stacked in series. Each PC is composed of materials a and b, each of thickness d_a and d_b arranged in a periodic manner. The PC on top (denoted by PC1) has 6 periods whereas the PC at the bottom (denoted by PC2) has 30 periods. The short PC1 allows the incident wave to couple to the interfacial acoustic wave while the long PC2 acts like a semi-infinite solid that does allow the penetration of wave into great depth.

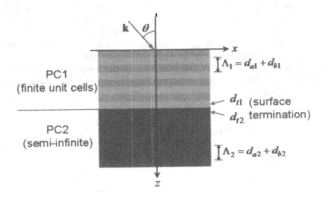

Figure 2. Phononic crystal (PC) design. PC1 consists of 6 periods and PC2 consists of 30 periods.

The elastic wave fields are calculated by Kennett's method [10], which remedies numerical problems at high frequencies of an earlier method using the transfer matrix. In the transfer matrix method, upgoing and downgoing waves are propagated (transferred) through every layer of different materials by performing matrix multiplications. The transfer matrices contain exponentially growing terms that cancel each other through the various matrix operations, thus causing numerical problems especially at high frequencies. Kennett reformulated the solution method by deriving recursive relations that completely eliminates the exponentially growing terms in the equations, thus avoiding numerical noise at high frequencies.

In PC1, the densities and the Lame's constants of materials a and b are $\rho_a = 1050$ kg/m^3, $\rho_b = 1380$ kg/m^3, $\lambda_a = 0.96$ GPa, $\lambda_b = 3.12$ GPa, $\mu_a = 0.96$ GPa, $\mu_b = 1.04$ GPa, and the materials properties of PC2 are $\rho_a = 1100$ kg/m^3, $\rho_b = 1300$ kg/m^3, $\lambda_a = 0.7857$ GPa, $\lambda_b = 2.925$ GPa, $\mu_a = 1.0$ GPa, and $\mu_b = 1.3$ GPa. The incident medium was set to be water with density $\rho = 1050$ kg/m^3, the first and second Lame's coefficients are $\lambda = 2.36$ GPa and $\mu = 0$ GPa. In order to realize the absorption, 0.1% of losses were added into λ and μ for the constituent materials such that the first Lame's coefficient, for instance, becomes a complex quantity after adding losses as $\tilde{\lambda} = \lambda(1 - 0.001i)$ with $i = \sqrt{-1}$. Here the imaginary part has to be negative in order for the propagating wave to decay in the medium. On the other hand, water is regarded as a lossless medium. For simplicity, the elastic properties and losses are assumed to be independent of the frequency and we set $d_a = d_b = \Lambda / 2$ for both PCs.

RESULTS

Figure 3 shows the calculated reflectance at the surface of the proposed structure due to a pressure wave incident at normal incidence. The reflectance is calculated as the ratio of incident energy to the reflected energy based on the energy flux of acoustic waves [17]. In Fig. 3, the top panel illustrates the reflectance of the PC1 with 6 unit cells submerged in water. As can be seen from the reflectance spectrum, there exists a stop band where energy-carrier wave cannot propagate into the medium, that is, the reflectance is closed to unity. Since 6-period PC1 with 0.1% loss is used in the calculation, the reflectance in the stop band cannot become unity and the stop-band edges are not sharp. However, the stop band in the reflectance spectrum appears to be much shaper in the middle panel for the 30-period PC2. In general, the stop bands can be easily shifted by changing the period.

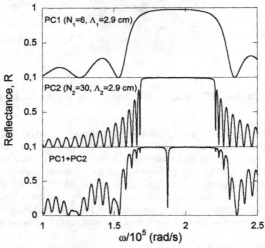

Figure 3. Surface reflectance. (top panel) PC1, (middle panel) PC2, and (bottom panel) PC1 + PC2

As shown in the bottom panel in Fig. 3, when PC1 is stacked above PC2 such that the pressure wave is incident on the PC1, there occurs a sharp reflectance dip near the frequency of 1.87×10^5 rad/s that is in the stop band of the individual PCs. Since PC2 is semi-infinite, the reduction of incident energy is eventually absorbed by the structure. This enhanced absorption in the stop band is due to the excitation of the interfacial acoustic wave, believed to be the Stoneley wave [18]. Similar to the electromagnetic surface waves, the incident energy is resonantly coupled to the interfacial acoustic wave and absorbed by the layers near the interface between two PCs. By considering 0.1% losses in the constituent layers of PCs, we can achieve approximately 90% absorption of the incident energy when the surface termination is $d_{t1} = d_{t2} = 0.05\Lambda$.

CONCLUSION

In summary, we have shown that interfacial acoustic waves can be excited at the interface between two PCs in the frequency regions where the stop band of two PCs overlaps. The

196

interfacial acoustic wave existing between two PCs is analogous to the surface electromagnetic wave existing between two photonic crystals. By excitation of interfacial acoustic waves, nearly 90% absorption can be achieved even at the normal incidence. The results obtained from the present study will facilitate the design of mechanical wave absorbers and acoustic metamaterials for military and engineering applications.

ACKNOWLEDGEMENT

The support for this research from the Swanson School of Engineering of the University of Pittsburgh is gratefully acknowledged.

REFERENCES

1. S. A. Cummer and D. Schurig, New J. Phys., **9**, 45 (2007).
2. H. Chen and C. T. Chan, Appl. Phys. Lett., **91**, 183518 (2007).
3. D. Torrent and J. Sánchez-Dehesa, New J. Phys., **9**, 323 (2007).
4. F. Villa and J. A. Gaspar-Armenta, Opt. Commun., **223**, 109 (2003).
5. S. Gonella, A. C. To, and W. K. Liu, J. Phys. Mech. Solid. (in press).
6. S. Gonella and M. Ruzzene, J. Sound Vib. **312**, 125 (2008).
7. B. Djafari-Rouhani, L. Dobrzynski, and O. Hardouin Duparc, Phys. Rev. B, **28**, 1711 (1983).
8. R. E. Camley, B. Djafari-Rouhani, L. Dobrzynski, and A. A. Maradudin, Phys. Rev. B, **27**, 7318 (1983).
9. G. Wang, D. Yu, J. Wen, Y. Liu, X. Wen, Phys. Lett. A, **327**, 512 (2004).
10. B. L. N. Kennett, *Seismic Wave Propagation in Stratified Media* (Cambridge University Press, 1983).
11. P. Yeh, *Optical Waves in Layered Media* (Wiley, New York, 1998).
12. H. Raether, *Surface Plasmons on Smooth and Rough Surfaces and on Gratings* (Springer-Verlag, 1998).
13. K. Park, B. J. Lee, C. J. Fu, and Z. M. Zhang, *J. Opt. Soc. Am. B*, **22**, 1016 (2005).
14. B. J. Lee, C. J. Fu, and Z. M. Zhang, Appl. Phys. Lett., **87**, 071904 (2005).
15. B. J. Lee and Z. M. Zhang, J. Appl. Phys., **100**, 063529 (2006).
16. B. J. Lee, Y.-B. Chen, and Z. M. Zhang, Opt. Lett., **33**, 204 (2008).
17. B. A. Auld, *Acoustic Fields and Waves in Solids*, Volume 1 (Krieger Publishing Company, Florida, 1990).
18. K. Aki and P. G. Richards, *Quantitative Seismology*, 2nd Edition (University Science Publishing, New York, 2002).

Mater. Res. Soc. Symp. Proc. Vol. 1188 © 2009 Materials Research Society   1188-LL09-03

Hydrophobic Metallic Nanorods Coated With Teflon Nanopatches by Glancing Angle Deposition

Wisam J. Khudhayer, Rajesh Sharma, and Tansel Karabacak

Department of Applied Science, University of Arkansas at Little Rock, AR, 72204

ABSTRACT

Introducing a hydrophobic property to vertically aligned hydrophilic metallic nanorods was investigated experimentally and theoretically. First, platinum nanorod arrays were deposited on flat silicon substrates using a sputter Glancing Angle Deposition Technique (GLAD). Then a thin layer of Teflon (nanopatches) was partially deposited on the tips of platinum nanorod at a glancing angle of $\theta = 85°$ as well as at normal incidence ($\theta = 0°$) for different deposition times. We show that GLAD technique is capable of depositing ultrathin isolated Teflon nanopatches on selective regions of nanorod arrays due to the shadowing effect during GLAD. Contact angle measurements on Pt/Teflon nano-composite have shown contact angle values as high as 138°, indicating a significant increase in the hydrophobicity of originally hydrophilic Pt nanostructures. Finally, a 2D simplified wetting model utilizing Cassie and Baxter theory of heterogeneous surfaces has been developed to explain the wetting behavior of Pt/Teflon nanocomposite.

INTRODUCTION

Metallic nanostructures have attracted great attention due to their novel properties which are of high interest in many applications such as hydrogen production and storage, surface catalysis, and heat transfer, etc. [1]. Recently, the hydrophilicty of vertically aligned metal nanorods with sharp nanotips was investigated experimentally [1]. It was found that as the surface roughness increases, the contact angle of the metallic nanorods decreases and therefore these nanostructures resulted in hydrophilic surfaces. On the other hand, many studies have focused on utilizing Radio Frequency (RF) sputtering technique to deposit Polytetrafluoroethylene (PTFE), commonly known as Teflon, on rough surfaces to get superhydrophobic surfaces [2-8]. It has been documented that an increase in PTFE film surface roughness increases the contact angle of water and therefore hydrophobicity without altering the surface chemistry [2,3]. In these studies, the resulting Teflon coating was a continuous rough film that enhanced the hydrophobicity.

However, as the micro- and nano-technology based systems emerge, conventional continuous PTFE coatings that completely cover the underlying surface may not be suitable for these applications as they block the desired transfer of photons/atoms/particles from/to the outside environment. Therefore, some applications may require "hydrophobic yet still isolated not-fully-coated nanostructured surfaces". To the best of our knowledge, there is no reported study on controlling the hydrophilicity of metallic nanorods with nanoscale roughness, even though there are many important applications for such nanostructures in surface catalysis, hydrogen production/storage, and heat transfer.

In this work, a novel glancing angle deposition (GLAD) technique was used to deposit ultrathin isolated Teflon nanopatches selectively on the tips of platinum (Pt) nanorods. GLAD technique provides a novel capability for growing 3D nanostructure arrays with interesting material properties [9-11]. It is a simple and single-step process unlike the surface roughening and surface modification approaches mentioned above. The GLAD technique uses the "shadowing effect," which is a "physical self-assembly" process through which obliquely incident atoms/molecules can only deposit to the tops of higher surface points, such as to the tips of a nanostructured array or to the hill-tops of a rough or patterned substrate. We show that the contact angle of the composite structure of Pt nanorods with Teflon nanopatches at the tips dramatically increases from hydrophilic values of uncoated nanorods to the highly hydrophobic values after coating with Teflon tips.

EXPERIMENT

In our experiments, a DC magnetron sputtering system was employed for the fabrication of Pt nanorod arrays. The depositions were performed on native oxide p-Si (100) wafer pieces (substrates size 3×3 cm^2), using a 99.99 % pure Pt cathode (diameter about 7.6 cm). The substrates were mounted on a sample holder located at a distance of about 18 cm from the cathode. They were tilted so that the angle between the surface normal of the target and the surface normal of the substrate was $\theta_{dep} = 85°$. The substrates were rotated around the surface normal with a speed of 30 RPM. The base pressure of about 4×10^{-7} Torr was achieved using a turbo-molecular pump backed by a mechanical pump. In all deposition experiments, the power was 200 Watts with an ultra pure Ar working gas pressure of 2.0×10^{-3} Torr. The substrate temperature during growth was below ~85 °C. The deposition time was 60 minutes. The deposition rate of the glancing angle depositions of Pt nanorods was measured to be about 10 nm/min from the analysis of cross- sectional SEM images.

After fabricating Pt nanrods, Teflon was deposited on top of Pt nanorods by utilizing an RF sputter deposition at a glancing angle of $\theta_{dep} = 83.7°$ (GLAD) for different deposition times of 1, 5, 15, and 30 minutes. For the normal incidence ($\theta_{dep} = 0°$), the deposition times were 20 seconds and 5 minutes. GLAD allows coating Teflon only to the tips of the Pt nanorods resulting in a bi-layer structure nanorod structure (Pt base and Teflon tip) while normal incidence results in a continuous Teflon thin film coating. A custom-made Teflon (Applied Plastics Technology, Inc.) disk was used as the sputtering target. The target was 0.3175 cm thick and 5.08 cm in diameter. The substrates (arrays of Pt nanorods on silicon wafer piece) were rotated around the surface normal with a speed of 1 RPM. The deposition was performed under a base pressure of about 4×10^{-7} Torr. During Teflon deposition experiments, the power was 150 Watts with an ultra pure Ar working gas pressure of 3.2×10^{-3} Torr. Finally, our composite (Pt/Teflon) structures were analyzed using scanning electron microscopy (SEM) and the hydrophobic behaviour was investigated by contact angle measurements using a VCA optima surface analysis system (AST Products, Inc., MA). In addition, elemental chemical analysis on sample surface was made using an energy dispersive x-ray analysis (EDAX) system attached to the SEM unit.

200

RESULTS AND DISCUSSION

Scanning Electron Microscopy (SEM) was used to study the morphology of our multifunctional composite (Pt/Teflon) nanostructures. Figure 1 shows the SEM images of pure Pt nanorods and the composite structure of Pt nanorods with Teflon tips which are deposited using RF sputtering technique at a glancing angle as well as at normal incidence for different deposition times. It was challenging to get clear SEM images of Pt/Teflon composites due to the charging of Teflon surface. However, this charging helped us to locate the Teflon coated regions on the Pt nanorods, which was visualized as a whitish coating in cross sectional SEM images.

Based on SEM images analysis, it was found that GLAD technique was able to deposit Teflon selectively on the tips of Pt nanorods, which results in isolated arrays of composite nanostructures. On the other hand, conventional normal incidence deposition of Teflon on Pt nanorods resulted in a continuous Teflon capping thin film layer lying mainly at the tips of Pt nanorods. We also observed that as the deposition time increases, Teflon islands tend to coalesce with other Teflon islands on neighboring nanorods in both normal incidence and GLAD depositions, which results in a smoother Teflon surface at the top and a decrease in the contact angle values. In general, for normal angle deposition, coalescence of Teflon islands is more pronounced, film quickly gets smoother, and therefore contact angle values decreases faster compared to the GLAD Teflon as shown in Table 1).

Figure 1: Top and cross-section views of (a) bare glancing angle deposition (GLAD) Pt nanorods, (b) Pt nanorods with GLAD-Teflon nanopatches at the tips are shown (ultrathin Teflon deposition was made for 1 minute).

Table 1: Measured contact angle values for various Teflon deposition time and Teflon thickness using either normal incidence (capping) or glancing angle deposition (GLAD-nanopatches) technique are listed.

Sample number	Sputtering Mode	Deposition time	Teflon Thickness (nm)	Contact angle
1	Normal Incidence (capping)	20 s	4.26	130°
2	Normal Incidence (capping)	5 minutes	64	122°
3	GLAD nanopatches	1 minute	4.27	138°
4	GLAD nanopatches	5 minutes	21.35	135°
5	GLAD nanopatches	15 minutes	64	133°
6	GLAD nanopatches	30 minutes	128	132°

Energy Dispersive X-ray Analysis (EDAX) was also utilized for elemental analysis and mapping of Pt/Teflon composites (samples 2 and 4 in Table 1). EDAX analysis (not shown) reveals that the elements present in our composite samples are carbon, fluorine, Pt, and silicon. True carbon to fluorine ratio for a chemical composition analysis of Teflon layer cannot be determined from EDAX plots due to the carbon contamination in EDAX chamber. In addition, we also analyzed the spatial EDAX distribution of fluorine atoms (not shown) mapped for GLAD and normal incidence deposited Teflon on Pt nanorods. In this analysis, although the fluorine atoms boundaries were not well defined due to the size of EDAX beam, which is about 100 nm, it could be seen that the density of fluorine at the tips of Pt nanorods is higher than that are at the gaps. This indicates that Teflon is concentrated on the tips of Pt nanorods when it is deposited by GLAD. This result further supports our SEM image analysis and shows that GLAD is capable of producing isolated composite nanostructures. On the other hand, for normal incidence deposition of Teflon, the distribution of fluorine atoms was relatively more uniform on the Pt nanorods and in the gaps compared to the GLAD Teflon nanopatches.

Contact angle measurements were performed for characterization of bare Pt nanorods, conventional flat Teflon thin film, Pt nanorods coated with normal incidence deposited Teflon film, and Pt-nanorods with ultrathin GLAD Teflon tips (nanopatches) using a VCA optima surface analysis system. In the literature, the term "ultrathin" films mean that the thickness of the films is less than about 5 nm [3]. Measured contact angle values of Pt nanorods, Teflon thin film, and Pt-nanorods-coated with Teflon tips are listed in table 1. It was found that the average contact angle of Pt nanorods was about 52° indicating a hydrophilic surface. Similarly, for the normal angle deposited flat Teflon thin film, the average contact angle was about 108° which indicates a hydrophobic surface, and it is in close agreement with the previously reported values of the contact angle of Teflon films [5]. As can be seen in Table 1, higher contact angle values of composite (Pt/Teflon) have been measured indicating a significant increase in the hydrophobicity of originally hydrophilic Pt nanostructures. This newly imparted hydrophobicity of nanorods may be attributed to the presence of low surface energy Teflon nanopatches with large surface area as can be observed in the SEM images shown in Fig.1.

In order to better understand the wetting of composite nanorods, a simplified two dimensional model has been developed utilizing Cassie and Baxter theory [12] of partial wetting of rough surfaces that leads to a heterogeneous interface formed by contacts of solid and vapor (air) with the liquid. In our model illustrated in Fig. 2a, d represents the water depth measured from the tip of the nanorods, a is the diameter of the nanorods, b is the gaps among the nanorods, t is the portion of Teflon at the side walls of the nanorods starting from the base line of the nanorods tips, and α is the tilt angle of the facets of the nanotips measured from the line parallel to the bottom plane. The average diameter of the nanorods is around 150 nm which measured from the SEM images. Under the Cassie and Baxter assumption, the fluid forms a composite surface with the solid where the water droplet sits upon a composite surface of the solid tops and the air gaps. Therefore, the Wenzel's model, which assumes that the fluid completely wets the solid structure, was modified by introducing the fractions f_s and f_a which correspond to the area in contact with the liquid and the area in contact with the trapped air beneath the drop, respectively [1,12]:

$$\cos \theta_{CB} = f_s \cos \theta_Y + f_s - 1 \qquad (1)$$

where, θ_Y is the contact angle that a liquid drop makes with an ideally flat surface (Young's theory) and f_s is the area fraction of the solid-fluid interface. As can be seen from Eq. (1) that if f_s tends to zero, the contact angle approaches 180° and as f_s tends to one, the expression tends to the Wenzel's equation. In our model, the nanostructured surface is flattened so that the water droplet sits upon a composite surface of the solid tops and the air gaps. This approximation is especially valid since the water droplet size is much bigger than the feature size of nanostructured surface as in the case of our experiments. Therefore, the Cassie and Baxter equation can be applied with two assumptions: first, assuming the shape of nanorods to be a cylinder with pyramidal tips; second, the average contact angle of a flat surface composed of Teflon and Pt portions, where the water completely wets (i.e. no air gaps) is given by:

$$\cos \theta_{Y_{Pt-Teflon}} = f_t \cos \theta_t + f_{Pt} \cos \theta_{Pt} \tag{2}$$

where, f_t and f_{Pt} are the area fractions of both Teflon and Pt in touch with water and θ_t and θ_{Pt} are the contact angles of flat Teflon and flat Pt surfaces, respectively.

When the water is partially wetting the Pt/Teflon nanorods and there exists air gaps at the bottom between the nanorods, then Pt/Teflon portion that is in contact with the water will contribute the solid fraction term f_s in Eq. (1). And, the term θ_Y will be replaced by the final average contact angle of the wetted portion of the Pt/Teflon surface. Therefore, using Eq. (1) and (2), modified Cassie and Baxter equation for our composite nanorods become:

$$\cos \theta_{CB_{Composite}} = f_{s_{composite}} \cos \theta_{Y_{Pt-Teflon}} + f_{s_{composite}} - 1 \tag{3}$$

where, $f_{scomposite}$ is the area fraction of solid-liquid (Pt/Teflon portion in contact with water) interfaces. Equation (1) can be used when the water is wetting Teflon only. On the other hand, when the water is wetting both Pt and Teflon, the Equation (3) can be applied. The fraction of solid-water interface can also be represented in terms of water depth d (Fig. 2a) penetrating into the gaps of nanorods as measured from their tips (e.g. no wetting when water depth is zero and complete wetting when it is equal to nanorods length):

$$f_s = \frac{2d / \sin \alpha}{[(2d / \sin \alpha) + ((a \tan \alpha - 2d) / \sin \alpha) + b]} \tag{4}$$

$$f_{s_{composite}} = \frac{[((2 \times (a/2) \times \tan \alpha) / \sin \alpha) + d - ((a/2) \times \tan \alpha)]}{[((2 \times (a/2) \times \tan \alpha) / \sin \alpha) + d - ((a/2) \times \tan \alpha) + b]} \tag{5}$$

$$f_{s_{composite}} = \frac{[((2 \times (a/2) \times \tan \alpha) / \sin \alpha) + (d - t - ((a/2) \times \tan \alpha))]}{[((2 \times (a/2) \times \tan \alpha) / \sin \alpha) + (d - t - ((a/2) \times \tan \alpha)) + b]} \tag{6}$$

In our model, the water depth d has been changed in a wide range of values in order to predict the contact angle of Pt/Teflon composite at different values of d. Hence, different scenarios can be considered: first, Teflon is only at the pyramidal tips of the nanorods and water is partially wetting Teflon only. Therefore, Eq. (4) can be applied followed by Eq. (1) to determine the contact angle. In the second case, water is completely wetting Teflon tips and also partially in contact with Pt base. Hence, after calculating the $f_{scomposite}$ value from Eq. (5), the contact angle of Pt/Teflon composite can be calculated using Eq. (3). And finally, the third scenario assumes that the Teflon, which completely covers the tips of Pt nanorods, is also assumed to partially coat the upper portion of the Pt side walls at bottom of tips due to the flux angular distribution effect explained above. Similarly, the contact angle in this case can be calculated from Eq. (3) in which the $f_{scomposite}$ value can be extracted from Eq. (6).

Since there is a possibility that an unknown amount of Teflon might have been deposited at the side walls of Pt nanorods due to the angular distribution in the sputter flux in our experiments, we also studied the effect of Teflon side wall coating portion (t in Fig. 2a) on the contact angle of Pt/Teflon nano-composite. For this we changed the parameter t in Eq. (6) for nanorods with tip angle $\alpha = 45°$, nanorod diameter a = 150 nm, nanorod gap $b = 50$ nm, plotted predicted contact angle values in Fig. 2b. The result in Fig. 2b shows that as Teflon side wall portion increases, the contact angle increase for a given water depth. This is due to fact that water is in contact with more Teflon for large value of t compared to the case where Teflon just coated the tips of Pt nanorods. As presented above, our experimental contact angle of Pt/Teflon composite is about 138° for nanorod coated with GLAD-Teflon. Therefore, according to the result of our model plotted in Fig. 2b, it is predicted that solid-liquid interface is expected to be mainly at Teflon tips when the composite nanorods are in contact with water.

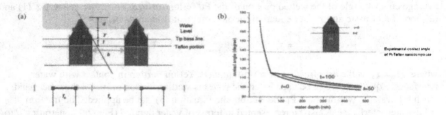

Figure 2: (a) Cross section of the simplified wetting model on Pt/Teflon nanocomposite. The composite surface is flattened so that the Cassie and Baxter theory can be applied to predict the composite contact angle and (b) Contact angle values as a function of water penetration depth, as predicted by our wetting model for various values of the Teflon portion at the Pt nanorods side walls t (apart from the Teflon at the tips).

CONCLUSIONS

The wetting of water on a composite nanostructured surface formed by arrays of Pt with Teflon nanopatches has been investigated experimentally and theoretically. A sputter glancing angle deposition (GLAD) technique was used to fabricate nano-composite Pt/Teflon surface. We demonstrated that the hydrophilic Pt nanostructured surfaces can be turned into highly hydrophobic surfaces by adding a little amount of Teflon at the tips of Pt nanorods. The contact angle measurements on this composite have shown contact angle values as high as 138°, indicating a significant increase in the hydrophobicity as compared to the originally hydrophilic Pt nanostructures with contact angle value about 52°. In addition, a 2D simplified wetting model utilizing Cassie-Baxter theory of heterogeneous surfaces has been developed to explain the wetting behavior of Pt/Teflon nanocomposite.

The authors would like to thank to the UALR Nanotechnology Center staff Dr. Fumiya Watanabe for his valuable support and discussions during SEM measurements.

REFERENCES

1. D. –X. Ye, T.-M Lu, and T. Karabacak, Physical review letters, PRL **100**, pp. 256102, 2008.
2. Satyaprasad, V. Jain, and S. K. Nema, Applied Surface Science, **253**, pp. 5462-5466, 2007.
3. D. K. Sakar, M. Farzaneh, and R. W. Paynter, Material Letters, **62**, pp. 1226-1229, 2008.
4. H. Biederman, Vacuum, **59**, pp. 594-599, 2000.
5. H. Biederman, M. Zeuner, J. Bilkova, et al., Thin Solid Films, **392**, pp. 208-213, 2001.
6. H. Biederman, V. Stelmashuk, I. Kholodkov, et al., Surface Coatings Technology, **174-175**, pp. 27-32, 2003.
7. S. Pursel, NNIN REU Research Accomplishments, pp. 104-105, 2004.
8. D. S. Bodas, A. B. Mandale, and S. A. Gangal, Applied Surface Science, 245, pp. **202-207**, 2005.
9. K. Robbie, G. Beydaghyan, T. Brown, C. Dean, J. Adams, and C. Buzea, Rev. Sci. Instrum. **75**, 1089 (2004).
10. T. Karabacak and T.-M. Lu, *Handbook of Theoretical and Computational Nanotechnology*, edited by M. Rieth and W. Schommers (American. Scientific Publishers, Stevenson Ranch, CA, 2005), chap. **69**, p. 729.
11. T. Karabacak, G. C. Wang, and T.-M. Lu, J. Vac. Sci. Technol. **A 22**, 1778 (2004).
12. D. M. Spori, T. Drobek, S. Zurcher et al., Langmuir, **24**, pp. 5411-5417, 2008.

Mater. Res. Soc. Symp. Proc. Vol. 1188 © 2009 Materials Research Society

Sound Absorption Characteristics of Porous Steel Manufactured by Lost Carbonate Sintering

Miao Lu[1], Carl Hopkins[2], Yuyuan Zhao[1], Gary Seiffert[2]
[1]Department of Engineering, University of Liverpool, Liverpool, L69 3GH, UK
[2]School of Architecture, University of Liverpool, Liverpool, L69 3BX, UK

ABSTRACT

This paper investigates the sound absorption characteristics of porous steel samples manufactured by Lost Carbonate Sintering. Measurements of the normal incidence sound absorption coefficient were made using an impedance tube for single-layer porous steel discs and assemblies comprising four layers of porous steel discs. The sound absorption coefficient was found not to vary significantly with pore size in the range of 250-1500 µm. In general, the absorption coefficient increases with increasing frequency and increasing thickness, and peaks at specific frequencies depending on the porosity. An increase in porosity tends to increase the frequency at which the sound absorption coefficient reaches this peak. An advantage was found in using an assembly of samples with gradient porosities of 75%-70%-65%-60% as it gave higher and more uniform sound absorption coefficients than an assembly with porosities of 75%.

INTRODUCTION

Porous metals have multifunctional properties. They retain some properties of metals such as good electrical and thermal conductivity, and also possess the special characteristics of porous structures such as good energy absorption and sound absorption. Recently, there is considerable interest in the study of the properties of porous metallic materials, especially their sound absorption. Porous metals are particularly suitable for use under extreme conditions where high temperatures, high noise levels, high air velocity, and high humidity may exist.

The capability of a material to absorb sound is measured by sound absorption coefficient which is defined as the ratio of the absorbed sound intensity to the incident sound intensity. Porous metals that are effective absorbers can have sound absorption coefficients exceeding 0.9 [1]. Sound absorption coefficient of porous materials is dependent on material properties, such as pore morphology, pore tortuosity, porosity, airflow resistivity and sample thickness, and sound frequency [2].

Lu *et al.* investigated the sound absorption of metal foams with open and closed cells and studied the effect of different processing methods on sound absorption coefficients [3, 4]. They found that the sound absorption properties of foams with closed cells after minor compression were poor because the largely closed cell structure prevents air particles penetrating inside the material. As a result, sound energy can not be dissipated within these porous materials. Han *et al.*

studied the sound absorption behavior of open-cell aluminum alloy foam manufactured by the infiltration method [5, 6]. They discovered that the sound absorption of the foam at frequencies over 1000Hz is better than the commercial metal foams available. They proposed that the complex internal pore structures and rough internal surfaces allowed more sound to be absorbed through viscous and thermal losses. Bo and Tianning studied the sound absorption coefficient of a porous sintered fiber metal [7]. They used a simple acoustic model to calculate the sound absorption coefficient, using convective heat transfer in metallic tube for reference. They found that the assembly order of porous fiber metal sheets and relative density can have significant effects on the sound absorption coefficient.

This paper studies the sound absorption characteristics of open-cell porous steel manufactured by the Lost Carbonate Sintering (LCS) process developed by Zhao *et al.* [8]. The objective is to investigate the effects of porosity, pore size and thickness of the porous steel in sound absorbing process by comparing the sound absorption coefficients. It also studies the effect of assemblies of porous steel samples with different porosities.

EXPERIMENTAL

The porous steel samples were produced by the LCS method [8, 9]. Astaloy A steel powder was first mixed with potassium carbonate granules with a specific size range at a specific volume ratio. The mixture was then compressed in a steel tube with an internal diameter of either 30 mm or 100 mm by a hydraulic press. The 30 mm and 100 mm preforms were then sintered in a furnace at 850°C for 5 and 8 hours, respectively, and air cooled to room temperature. The sintered preforms were machined to desired thicknesses and their faces were polished while potassium carbonate granules were still in the samples. The potassium carbonate was then dissolved by hot water, resulting in circular porous steel discs. The samples have a diameter of 30 or 100 mm; a thickness of 5, 10 or 20 mm; a pore size range of 250-425, 425-710, 710-1000 or 1000-1500 μm; and a nominal porosity of 60, 65, 70 or 75%. The actual porosity measured by the Archimedes method is slightly higher (<2%) than the nominal porosity.

The normal incidence sound absorption coefficient of the porous steel samples was measured using the impedance tube method using samples on a rigid backing. A comprehensive description of the method and the transfer-function calculation can be found in EN ISO 10534-2:2001 [10]. The measurement equipment consisted of an impedance tube, a test sample holder, two microphones, a signal generator, a two-channel Fast Fourier Transform (FFT) analyser, a power amplifier and a loudspeaker, as shown schematically in Figure 1. The diameter of the impedance tube determines the valid frequency range of the result. Different diameter samples and tubes were used for different frequency ranges. For the frequency range from 50 to 500Hz, the samples had diameters of 100 mm and diameters of 30 mm for 500 to 5000Hz. Vaseline was used around the perimeter of the samples to ensure that there were no air gaps between the tube and the samples.

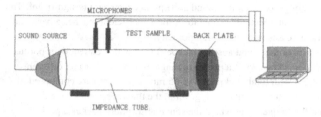

Figure 1. Schematic diagram showing the impedance tube measurement.

RESULTS

Figure 2 shows the variation of the sound absorption coefficient with frequency for 10 mm thick porous steel samples with (a) a fixed nominal porosity of 70% and different pore size ranges and (b) a fixed pore size range of 425-710 µm for different porosities between 60% and 75%. For 50 to 500Hz, the sound absorption coefficients are below 0.2, but above this the sound absorption coefficients tend to increase rapidly with increasing frequency. For any porous layer on a rigid backing there are peaks of high absorption at specific frequencies. For the samples with a nominally identical porosity of 70%, the ranges of pore size do not show significant differences in the sound absorption coefficient (see Figure 2a).

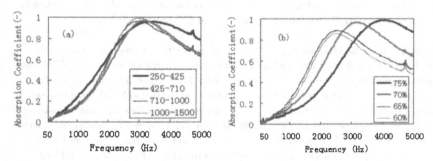

Figure 2. Effects of pore size and porosity on the sound absorption coefficient of porous steel samples with (a) different pore sizes (µm) and a fixed nominal porosity of 70%, and (b) different porosities and a fixed pore size range of 425-710 µm.

Figure 2b shows that the effect of porosity on the sound absorption coefficient is significant. The porosity determines the frequency at which the sound absorption coefficient reaches a peak and the peak frequency generally increases with increasing porosity.

Figure 3 shows the variation of the sound absorption coefficient with frequency for porous steel samples with a nominal porosity of 75%, a pore size range of 425-710 µm and different sample thicknesses. The samples with total thicknesses of 15 mm and 40 mm are composed of two layers, 10+5 mm and 20+20 mm respectively. In general, the thicker samples have the highest sound absorption and the frequency at which the sound absorption reaches a peak tends to decrease with increasing sample thickness.

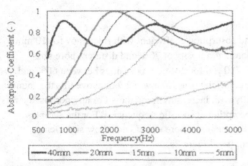

Figure 3. Effect of sample thickness on the sound absorption coefficient of porous steel samples with a nominal porosity of 75% and a pore size range of 425-710 µm.

Figure 4 shows the variation in the sound absorption coefficient with frequency for two assemblies of 10 mm thick porous steel samples with (a) the same nominal porosity of 75% and (b) different porosities of 75%-70%- 65%-60%, with the highest porosity sample facing the incident sound wave. All the samples have a fixed pore size range of 425-710 µm. It is seen that the 75%-70%-65%-60% sample assembly has sound absorption coefficients greater than 0.8 over a wide range of frequencies. In addition, the fluctuations with frequency are much lower than with the 75%-75%-75%-75% assembly. This indicates that there are potential advantages in using assemblies with a graduated decrease in porosity from the surface in order to achieve uniformly high sound absorption.

Figure 5 shows the variation of sound absorption coefficient with frequency for four assemblies of porous steel samples with gradient porosities of 75%-70%-65%-60%, all with a pore size range of 425-710 µm. The assemblies consist of 4 layers, 4×5, 4×10, 4×(10+5) and 4×20 mm respectively. It is shown that when the overall thickness is greater than 20 mm, the sound absorption coefficients tend not to fluctuate. In addition it is seen that at thicknesses above

40 mm the sound absorption coefficients do not increase very much with further increase in assembly thickness.

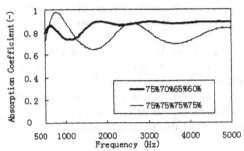

Figure 4. Effect of porosity gradient on the sound absorption coefficient of porous steel samples with a pore size range of 425-710μm.

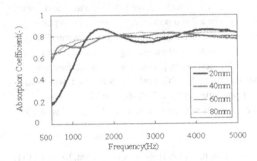

Figure 5. Effect of assembly thickness on sound absorption coefficient for porous steel samples with gradient porosity of 75%-70%-65%-60%.

DISCUSSION

The sound absorption of porous materials with a rigid frame mainly depends on the viscous resistance of air in the pores [11]. The friction between the vibrating air particles and the surface of the pores dissipates energy in the form of heat. Therefore, the main structural properties of the porous materials affecting sound absorption are pore size, porosity and thickness. In the range of pore size studied, 250-1500 μm, varying the pore size does not significantly affect the sound absorption coefficient. However, the porosity has a significant effect on sound absorption. High porosity porous steel samples have higher sound absorption coefficients at higher frequencies, whereas low porosity samples have higher sound absorption coefficients at lower frequencies. Thick samples also tend to have higher sound absorption, especially at low frequencies, as seen in Figure 4. The frequency of peak absorption coefficient decreases with increasing thickness. Overall, the sound absorption of porous steel is generally poor at low frequencies but increases significantly at high frequencies.

As the porosity determines the frequency of peak sound absorption coefficient, it is expected that a sample with a gradient porosity would have more uniform sound absorption over a broader frequency range than a sample of a single porosity. Comparison of the 75%-70%-65%-60% assembly with the 75%-75%-75%-75% assembly shows that the large fluctuations with the latter do not occur with the former. The result indicates that assemblies of samples with different porosities can be used to give a more uniform, high sound absorption coefficient.

CONCLUSION

The normal incidence sound absorption coefficient of porous steel was measured using an impedance tube. The sound absorption generally increases with increasing frequency, and peaks at specific frequencies depending on the porosity. For a fixed porosity, the sound absorption coefficient was found not to vary with pore sizes in the range of 250-1500 μm. In general, the absorption coefficient increases with increasing frequency and increasing thickness, and peaks at specific frequencies depending on the porosity. An increase in porosity tends to increase the frequency at which the sound absorption coefficient reaches this peak. An advantage was found in using an assembly of samples with gradient porosities of 75%-70%-65%-60% as it gave higher and more uniform sound absorption coefficients than an assembly with porosities of 75%.

REFERENCES
1. M.F. Ashby, *Metal Foams: A Design Guide*, (Butterworth Heinemann, Boston,2000) p. 171.
2. P.M. Morse, *Vibration and Sound*, (McGraw-Hill, New York, 1948).
3. T.J. Lu, F. Chen and D.P. He, J. Acoust. Soc. Am. **108**, 1697 (2000).
4. T.J. Lu, Acta Mech. Sin. **18**, 457 (2002).
5. F.S. Han, Z. Zhu and C Liu, Acta Acoust. **84**, 573 (1998).
6. F.S. Han, G. Seiffert, Y.Y. Zhao and B. Gibbs, J. Phys. D: Appl. Phys. **36**, 294 (2004).
7. Z. Bo and C. Tianning, Appl. Acoust. **70**, 337 (2009).
8. Y.Y. Zhao, T. Fung, L.P. Zhang and F.L. Zhang, Scrip. Mater. **52**, 295 (2005).
9. L.P. Zhang and Y.Y. Zhao, J. Eng. Manufacture, **222**, 267 (2008).
10. BS EN ISO 10534-2:2001, Acoustics-Determination of sound absorption coefficient and impedance in impedance tubes, Part2: transfer-function method.
11. L.J. Gibson and M.F. Ashby, *Cellular Solids: Structure and Properties*, (Cambridge University Press, Cambridge, 1997) p. 303.

Mater. Res. Soc. Symp. Proc. Vol. 1188 © 2009 Materials Research Society   1188-LL04-07

Heat Transfer Performance of Porous Copper Fabricated by the Lost Carbonate Sintering Process

Liping Zhang[1], David Mullen[2], Kevin Lynn[2] and Yuyuan Zhao[1]
[1]Department of Engineering, University of Liverpool, Liverpool L69 3GH, UK
[2]Thermacore Europe Ltd, Ashington, Northumberland NE63 8QW, UK

ABSTRACT

The heat transfer coefficients of porous copper fabricated by the lost carbonate sintering (LCS) process with porosity range from 57% to 82% and pore size from 150 to 1500 μm have been experimentally determined in this study. The sample was attached to the heat plate and assembled into a forced convection system using water as the coolant. The effectiveness of the heat removal from the heat plate through the porous copper-water system was tested under different water flow rates from 0.3 to 2.0 L/min and an input heat flux of 1.3 MW/m^2. Porosity has a large effect on the heat transfer performance and the optimum porosity was found to be around 62%. Pore size has a much less effect on the heat transfer performance compared to porosity. High water flow rates enhanced the heat transfer performance for all the samples.

Key words: Porous media, heat transfer, porous copper, LCS

INTRODUCTION

Porous metals with open cells have many potential applications in thermal management because of their huge internal surface area and high permeability for fluids. One such application is as heat exchangers in the cooling systems for electronic devices [1-3]. The cooling system is composed of the porous metal medium and a gas or liquid coolant flowing through its internal channels. Porous copper is an ideal medium for use as heat exchangers because of the high thermal conductivity of copper. At present most commercial porous copper used as heat exchangers is produced by powder metallurgy through sintering copper particles together without using any fillers. This conventionally particle-sintered porous copper usually has a low and narrow porosity range well below 50% [4-6]. The particle-sintered bronze samples [4] with porosities from 40% to 46% can enhance the heat transfer performance of the system up to 15 times for water and up to 30 times for air in comparison with an empty channel. The porous copper fabricated by investment casting or by the powder metallurgy using polymer fillers has a high and narrow porosity range over 80%. Several investigations [1, 7, 8] reported that an increase in either porosity or pore size generally reduced the heat transfer coefficient although pore size had much less effect than porosity. The porous copper samples with high porosities from 88% to 94% can enhance the heat transfer about 17 times in comparison with an empty channel. These porous copper samples showed heat transfer coefficients 2 to 3 times of those of the FeCrAlY and Al foams with a similar porosity and a similar coolant flow rate.

Due to the limitations of the manufacturing methods, however, the current copper foams have a porosity either less than 50% or higher than 80%. The heat transfer performance of porous copper in the medium porosity range from 50% to 80% has hardly been reported. The porous copper fabricated by LCS can have a large range of porosity from 50% to 85% and various pore sizes [9]. The LCS porous copper has a very good sintered strength and a homogeneous structure and therefore is a promising candidate for heat transfer applications. This paper investigates the heat transfer performance of porous copper manufactured by LCS, with a porosity range from 57% to 82% and a pore size range from 150 to 1500 µm, under different water flow rates and the fixed input heat flux. A particle-sintered porous copper sample with a porosity of about 40% was also tested for comparison.

EXPERIMENTAL

Totally eight porous copper samples were fabricated by the LCS process, details of which were described in [9]. The raw materials were a commercially pure (99.9%) copper powder with a particle size range of 75 to 125 µm and a food grade potassium carbonate powder, which was divided into three size ranges: 150-250 µm, 425-710 µm and 1000-1500 µm. The process condition was: compaction pressure 200 MPa; sintering temperature 850 °C and sintering time 4 hrs. The porosity and pore size of all the samples are listed in Table I. Sample 9 was produced by sintering large copper particles, about 1000 µm, at 950°C for 2 hours without using any filler. The pore diameter of this sample was estimated as 400 to 600 µm based on the copper particle size used. It was included for comparison purposes. Figure 1 shows the SEM image of a typical LCS porous copper sample.

Table I. Porosity and pore size of porous copper samples used in testing

Sample Number	1	2	3	4	5	6	7	8	9
Nominal Porosity, %	80	75	70	70	70	65	60	55	40
Measured Porosity, %	82	76	72	71	72	68	62	57	40
Pore Size, µm	425-710	425-710	425-710	150-250	1000-1500	425-710	425-710	425-710	400-600

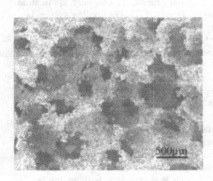

Figure 1 SEM image of a typical porous Cu sample produced by LCS

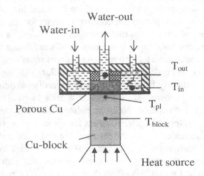

Figure 2 Schematic diagram of the heat exchange chamber used in the test

Figure 2 is a cross-sectional view of the testing chamber showing how the porous copper sample was mounted. The size of the samples is 10 mm in diameter and 4 mm in thickness. A heat plate was attached to the bottom of the sample; a constant input heat to the plate, depending on T_{block}- T_{pl}, was produced by a heat block heated by 5 heat cartridges. The water flow to the test chamber was produced by a CRI1S-27 pump, with a maximum pressure of 25 bar. The water entering the test chamber was kept cold by a Chiller. The temperatures of the heat block, heat plate, water inlet and outlet were measured by K and T type thermocouples. The water flow rate was measured by an ABB COPA-XL flow meter. All the data from the thermocouples and flow meter were recorded by a PC. The measurements were conducted after the steady state condition was reached for a constant input heat flux of 1.3 MW/m^2 and coolant flow rates from 0.3 to 2.0 L/min.

The heat transfer performance of the porous copper sample can be characterised by an overall heat transfer coefficient of the cooling system composed of the porous copper sample and the water flow. This heat transfer coefficient, h, can be determined by:

$$J = hA(T_{pl} - T_m) \tag{1}$$

where J is the heat flow from the heat plate to the cooling water; A is the contact area between the porous copper sample and the heat plate (A=78.5 mm^2 in this study); T_{pl} is the temperature of the heat plate; T_m is the mean temperature of the water inlet and outlet. If heat losses through the testing chamber are negligible, the heat flow, J, is equal to the amount of heat carried away by the cooling water in unit time, i.e.:

$$J = \rho C_p f (T_{out} - T_{in}) \tag{2}$$

where ρ is the density of water ($\rho = 1000$ kg/m^3); C_p is specific heat of water ($C_p = 0.6$ kJ /kgK); f is flow rate, m^3/s; T_{in} and T_{out} are the temperatures of water inlet and outlet. Combining Equations (1) and (2), the heat transfer coefficient can be obtained by:

$$h = \frac{\rho C_p f (T_{out} - T_{in})}{A(T_{pl} - T_m)} \tag{3}$$

RESULTS AND DISCUSSION

Effect of porosity

Six porous copper samples with the same pore size of 425-710 μm but different porosities from 57 to 82%, as shown in Table I, were tested in this study at flow rates from 0.3 to 2.0 L/min and an input heat flux of 1.3 MW/m^2. For comparison purposes the empty testing chamber, i.e. without a porous copper sample, and the particle-sintered porous copper sample, Sample 9, were also tested. They are designated in the following graphs by their porosities, 100% and 40%, respectively.

The variations of heat transfer coefficient with porosity at different water flow rates are shown in Figure 3. The porosity clearly has a large effect on the heat transfer coefficient. With increasing the porosity, the heat transfer coefficient first increases until reaching the highest values at the porosity 62% and then generally decreases for all flow rates. The empty chamber

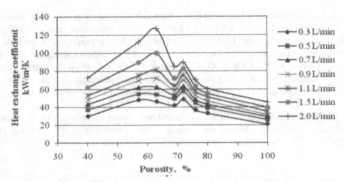

Figure 3. Effect of porosity on heat transfer coefficients of porous Cu samples at different water flow rates (pore size: 425-710 µm; input heat flux: 1.3 MW/m^2; compaction pressure: 200 MPa)

without a porous sample has the lowest heat transfer performance. It is evident that the porous copper sample has played an important role in the heat transfer of the system, increasing the heat transfer coefficient 2 to 3 times compared to the empty channel. Figure 3 also shows that the conventionally sintered copper without using a space holder, Sample 9, has a similar coefficient as Sample 2, which has a porosity of 76%. In other words, for the same heat exchange performance, the LCS porous copper has only 40% weight of conventionally particle-sintered copper. Comparing the LCS porous copper with a porosity of 62%, Sample 7, and Sample 9, the heat transfer coefficient of the former at different water flow rates are 1.5-1.8 times of those of the latter, while the weight of the former is only about two thirds of the latter. If the existing products are replaced by the LCS copper, then the heat exchangers would be much lighter or have better performance.

It is worth noting that there is a sharp fall in heat transfer coefficient when the porosity is increased from 62% to 68%. This is likely an outcome of a special structural feature of LCS porous copper, as this porosity coincides with the critical volume percentage of K_2CO_3 particles in a Cu-K_2CO_3 mixture where the interstices between the K_2CO_3 particles are just completely filled with smaller Cu particles. A previous study of the mechanical properties of the LCS porous copper showed that the porous copper with a porosity of 69% had lower impact strength than those of the porous copper with a slightly higher porosity [10], indicating that the structural integrity of the porous copper at a porosity around 68-69% is less good. Whether this is the right cause or not, however, needs further investigation.

The effect of porosity on the heat transfer performance can be analysed by its effects on the thermal conductivity, the internal surface area and the permeability of the porous sample. A high porosity corresponds to a low volume of copper matrix and thus a low thermal conductivity of the porous sample [11], which decreases the conductive heat transfer from the heat plate. At the same time a high porosity results in a high internal surface area and a high fluid permeability of the sample, which enhances the convective heat removal from the porous copper to the fluid flow. Therefore, the overall heat transfer performance depends not only on the thermal conduction in the copper matrix but also on the heat removal to the fluid. The optimum porosity is achieved by balancing these two processes. Low porosity is conducive to better thermal conduction from the heat source to the porous copper, but disadvantageous for convectional heat transfer from the porous copper to the coolant. In other words, thermal convection by the coolant is the limiting factor. High porosity has the opposite effect and the limiting factor is thermal conduction in the copper matrix. There exists an optimum porosity at which both conduction and convection are maximised. The optimum porosity resulting in best heat transfer found experimentally in this study is around 62%, as shown in Figure 3.

Effect of pore size

In order to investigate the effect of pore size on the heat transfer performance of the porous copper, three samples with different pore size ranges, 150-250 µm, 425-710 µm and 1000-1500 µm and a similar porosity of about 72% were tested. The heat transfer coefficients of the three samples at different water flow rates and the same input heat flux of 1.3 MW/m^2 are shown in Figure 4. The sample with the medium pore size range of 425-710 µm has heat transfer coefficients 1.5 times higher than those with the small and large pore size ranges. The small and large pore size samples have similar heat transfer performance. The effect of pore size on the heat transfer coefficient is considered to be an outcome of its effect on fluid permeability. For the same porosity, smaller pores have a higher specific surface area, leading to greater flow resistance. Larger pores, on the other hand, generally result in less straight or longer flow path. Both can lead to lower coolant permeability and thus lower convectional heat transfer. Compared with porosity, however, the effect of pore size is much less significant.

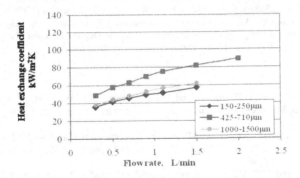

Figure 4. Effect of pore size on the heat transfer coefficients at different flow rates (porosity: about 72%, input heat flux: 1.3 MW/m^2, compaction pressure: 200 MPa)

Effect of water flow rate

The effect of water flow rate on the heat transfer coefficient of the porous copper samples with different porosities under the input heat flux of 1.3 MW/m^2 can also be seen in Figure 3. Increasing the flow rate from 0.3 to 2.0 L/min increases the coefficients of all the samples with different porosities. The samples with medium porosities (57% and 62%) show a large increase in coefficient with increasing the flow rate compared to the lower and higher porosity samples.

As expected from Equation (3), the heat transfer coefficient should increase nearly linearly with water flow rate for any fixed porosity. The effectiveness of increasing flow rate is clearly related to the role of convection in the overall heat transfer as discussed above.

CONCLUSIONS

Porous copper samples with large ranges of porosity and pore size have been fabricated by LCS and the heat transfer performance has been tested. Compared to the empty channel, introducing the porous copper sample can enhance the heat transfer performance 2 to 3 times. The porosity of the porous copper sample has a significant effect on the heat transfer performance, due to its large influence on the permeability and the heat conductivity of the sample. The optimum porosity, which balances these two factors, has been experimentally obtained as about 62% in this study. Pore size has a much less effect on the heat transfer coefficient. Increasing water flow rate increases the heat transfer performance for all samples, especially markedly for those with medium porosities, because of enhanced heat removal from the heat source. LCS porous copper has a large variable porosity range to achieve an optimum heat transfer performance. It is potentially a more effective material as heat exchangers for electronics cooling systems than the existing porous copper products.

ACKNOWLEDGEMENTS

This work is supported by a UK Technology Strategy Board project (TP/8/MAT/6/I/Q1568F). We would also like to thank Ecka Granules (UK) Ltd for supplying the copper powder and Mr Allan Ramsden at Thermacore Europe Ltd for his help in conducting the heat transfer tests. Liping Zhang would like to thank the Daphne Jackson Trust for a Fellowship supported by Equalitec Ltd and the Royal Academy of Engineering.

References

1. A.J. Fuller, T. Kim, H.P. Hodson and T. J. Lu, Proc. IMechE , **219**(2005), Part C: 183-191
2. G. Hetsroni, M Gurevich, R Rozenblit, Intl Journal of Heat and Fluid Flow, **27**(2006), 259-266
3. K. Boomsma, D. Poulikakos, and F. Zwick, Mechanics of Materials **35**(2003) 1161-1176
4. P.X. Jiang, M. Li, T.J. Lu, L.Yu and Z.P. Ren, Intl. Journal of Heat and Mass Transfer, **47**(2004), 2085-2096
5. S.C. Tzeng and W.P. Ma, Int. Comm. Heat Mass Transfer, **31**(2004), 827-836
6. G.J. Hwang and C.H. Chao, Journal of Heat Transfer, Transactions ASME, **116**(1994), 456-464
7. H.Y. Zhang, D. Pinjala, K. Joshi, T.N. Wong, K.C. Toh and M.K. Iyer, IEEE Transactions on Components and Packaging Technologies, **28** (2005), 272-279
8. C.Y. Zhao, T. Kim, T.J. Lu and H.P. Hodson, Journal of Thermophysics and Heat Transfer, **18**(2004), 309-317
9. Y.Y. Zhao, T. Fung, L.P. Zhang and F.L. Zhang, Scripta Materialia, **52**(2005), 295-298
10. X.F. Tao, L.P. Zhang and Y.Y. Zhao, Materials Science Forum, **539-543** (2007), 1863-1867
11. D.J. Thewsey and Y.Y. Zhao, Physica Status Solidi A, **205**(2008), 1126-1131

Mater. Res. Soc. Symp. Proc. Vol. 1188 © 2009 Materials Research Society   1188-LL07-02

Acoustic Properties of Organic/Inorganic Composite Aerogels

Winny Dong[a], Tanya Faltens[a], Michael Pantell[b], Diana Simon[c], Travis Thompson[d], and Wayland Dong[e]

California Polytechnic University, Pomona, CA, USA

  [a] *Chemical and Materials Engineering*

  [b] *Physics*

  [c] *Industrial Chemistry*

  [d] *Mechanical Engineering*

[e]*Veneklasen Associates, Santa Monica, CA, USA*

Abstract

Composite aerogels (with varying concentrations of silica and poly-dimethylsiloxane) were developed and their acoustic absorption coefficient as a function of composition and average pores size have been measured. The polydimethylsiloxane modified the ceramic structure of the silica aerogels, decreasing the material's rigidity while maintaining the high porosity of the aerogel structure. The composite aerogels were found to exhibit different modes of acoustic absorption than that of typical porous absorbers such as fiberglass. At some frequencies, the composite aerogels had 40% higher absorption than that of commercial fiberglass. Physical data show that these materials have a large surface area (> 400 m2/g) and varying pore sizes (d ~ 5 - 20 nm).

Introduction

When sound encounters a material, the pressure wave can be reflected, transmitted, or absorbed. Material properties such as elastic modulus and porosity directly influence how a material will interact with incident acoustic waves. For materials with a high elastic modulus, regardless of porosity, acoustic reflection is high. Materials with both a low modulus of elasticity and low porosity allow for acoustic transmission. In terms of absorption, the combination of low modulus and high porosity are ideal. (Table 1)

Table 1. Influence of elastic modulus and porosity on a material's acoustic response.

Modulus	Porosity	Example	Reflection	Transmission	Absorption
↑	↑	Concrete	**High**	Low	None
↑	↓	Steel	**High**	None	None
↓	↓	Water	Low	**High**	Low
↓	↑	Snow	Low	Low	**High**

In optimizing materials for architectural acoustic absorption, typically a high absorption coefficient across a wide range of frequencies is desired. Commercially available acoustic materials generally have high absorption at frequencies above 1000 Hz and negligible absorption below 800 Hz. The primary goal of this project is to develop materials with tunable modulus and porosity to prepare an adjustable acoustic absorber, specifically one that can absorb well below 800 Hz. A composite of silica aerogels and poly-dimethylsiloxane (PDMS) combines the low modulus and high porosity.

Silica aerogels have very high surface areas ($300 - 1000$ m^2/g) and high porosity ($80 -$ 99%) and can be synthesized through a liquid-precursor, room-temperature method. The high porosity leads to extremely low sound velocity (down to 90 m/s) through the silica aerogel. [1] However, the high elastic modulus also results in high reflection coefficients, especially at low frequencies. In order to modify the elastic modulus of the silica aerogels, PDMS is incorporated into the silica structure making Organically Modified Silicates (or ormosils). (Table 2) In contrast to traditional composites of alternating layers of mm-thick polymer and ceramic fibers, ormosils contain alternating polymer and ceramic groups at the molecular level giving a more homogeneous structure.

Table 2. It has been shown that the elastic moduli in silica/PDMS ormosils can be modified by controlling the concentration of PDMS. [2]

Mol % Polymer (PDMS)	0	7.9	12.8	19.5	23.7
Elastic Modulus (GPa)	**20.7**	**18.6**	**16.0**	**15.0**	**13.0**
Vickers Hardness (kg/mm^2)	186	160	140	110	88
Fracture toughness ($Mpa/m^{1/2}$)	0.50	0.49	0.48	0.47	0.46
Brittleness ($mm^{-1/2}$)	**3.63**	**3.19**	**2.86**	**2.32**	**1.88**

Additionally, the concentration of PDMS can modify the pore size of the ormosil aerogel. It has been shown that pore sizes of 8 nm and greater produce a Rayleigh scattering of phonons. [3] [4] Increasing the mean pore size has been shown to increase the sound attenuation of silica aerogels in addition to decreasing the longitudinal sound velocity. [4] The larger pore sizes also aid in better impedance matching with air, which leads to a lower reflectivity.

Experimental Procedures

The silica/PDMS ormosils were synthesized through combining 0.025 mol TEOS (Alfa Aesar), 0.0035 mol HCl, 0.042 mol THF (Acros Organics), 0.17 mol isopropanol (Acros Organics), 0.25 mol distilled water, and PDMS (Gelest Inc.). The PDMS:silica ratio varied between 0.05:0.95 to 0.40:0.60. The mixture was then sonicated (Branson Ultrasonics) for 60 minutes to achieve homogeneity. Gelation occurred within 20 minutes after sonication. The gels were allowed to age for 3 days and then washed with acetone to remove all liquid reaction byproducts. The acetone is then removed through CO_2 supercritical drying (Polaron ES300).

The absorption coefficients of the ormosil aerogels were measured in a home-made impedance tube, based on ASTM standards (C 384 04). The impedance tube is an enclosed tube with a speaker on one end and the acoustic absorber sample on the other end. An acoustic standing wave can be set-up within the tube and the wavelength and amplitude of this standing wave is measured with a movable microphone within the tube. By measuring the locations of the wave maximums and minimums, the Standing Wave Ratio (SWR) and the absorption coefficient (α) of the absorber can be calculated. The SWR is the ratio of the maximum acoustic pressure and the minimum acoustic pressure within the tube. The absorption coefficient represents the percentage of acoustic energy absorbed by the material.

$$\alpha = 1 - \frac{(SWR-1)^2}{(SWR+1)^2}$$

Surface area and average pore size were measured with nitrogen gas adsorption (Micromeritics, ASAP 2010). Scanning electron micrographs (SEM, Zeiss, Evo) were obtained to view microstructure.

Results and Discussion

Figure 1 shows the acoustic absorption profile of the silica/PDMS ormosils as a function of PDMS concentration. Their absorption profiles are compared to two commercially available acoustic insulators – fiberglass and a composite acoustic pad.

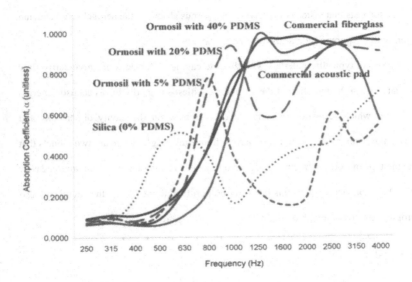

Figure 1. Adsorption coefficient as a function of the acoustic frequency. Silica/PDMS ormosils (with varying concentrations of PDMS) are compared with commercially available acoustic insulators.

There are three general types of acoustic absorption – porous absorbers, panel absorbers, and resonators. [5] The fiberglass and the acoustic pad exhibit typical porous absorber behavior. Their open cell structures allow air to flow through and the acoustic pressure wave is converted to heat as it passes through the material. This is the most common type of acoustic absorber with high α at frequencies above 1000 Hz and almost no absorption below 800 Hz. Common material characteristics considered by acousticians are open porosity (volume of free air per unit volume of porous medium) and surface area. [6] The silica aerogel (with no PDMS) and the ormosil sample with up to 20% PDMS exhibit very different absorption behaviors. The narrow absorption peak suggests that instead of behaving as a porous absorber, the aerogels are behaving as a resonator, with the pores acting as the resonating cavities.

Resonators are typically materials with discrete cavities. Air does not necessarily flow through the material but the vibration of the air in the cavities is what dissipates acoustic energy. The frequency at which resonators absorbs depend strongly on the cavity dimensions and therefore, resonators only absorb over a very narrow frequency range. There are two models that approximate the type of porous structure of the aerogels. The Helmholtz model assumes a cavity with a neck. The resonance frequency is inversely proportional to the size of the cavity and also inversely proportional to the length of that neck.

$$f_H = \frac{v}{2\pi}\sqrt{\frac{A}{VL}}$$

f_H = Helmholtz resonant frequency

v = speed of sound

A = cross-sectional area of neck

L = length of neck

V = volume of the cavity

If the cavities are shaped more like a sphere, the resonance frequency is inversely proportional to the diameter of the cavity and directly proportional to the size of the opening.

$$f = Y\sqrt{\frac{d}{D^3}}$$

F = resonant frequency

Y = material constant

d = diameter of the cavity opening

D = diameter of the cavity

In both cases, as the cavity size increases, resonance based absorption occurs at lower frequencies. [7]

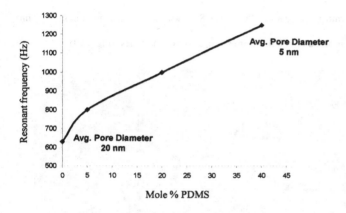

Figure 2. Resonant frequency as a function of PDMS concentration.

By increasing the PDMS concentration, the absorption peaks of the ormosils are shifted to higher resonant frequencies. This reflects the data that as the concentration of PDMS increases, the pore size of the ormosil aerogels decreases. As polymer is added, the average pore size decreases from 20 nm to 5 nm. (Table 3) Therefore, peak absorption frequency of the ormosil aerogels is inversely proportional to pore size, which matches the resonator model.

Mol % PDMS	Surface Area (m^2/g)	Avg. Pore Volume (cm^3/g)	Avg. Pore Diameter (nm)
0	800	8.0	20
40	450	0.5	5

Table 3. Surface area and average pore size of silica aerogels compared to an ormosil aerogel with 40% PDMS.

Scanning electron micrographs also confirm the initial pore size measurements. The SEMs of a 40% PDMS ormosil and that of a 20% PDMS ormosil look very similar at the 100 nm length scale. Both are composed of small, 10 nm sized, spherical particles. (Figure 3)

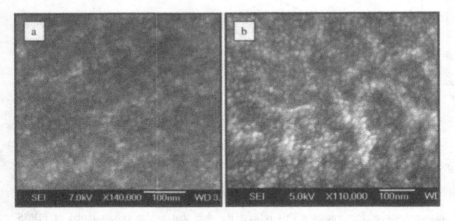

Figure 3. SEMs at the 100 nm-range of ormosil aerogels with a) 40% PDMS and b) 20% PDMS.

However, at the 10 um length scale the two materials look very differently. The 40% PDMS ormosil is much smoother with fewer pores whereas the ormosil with 20% PDMS has larger pores and a rougher structure. (Figure 4) The higher concentration of PDMS results in a smoother sample with fewer or smaller pores and the sample with less PDMS is more porous with larger pores. This further confirms that the ormosils exhibit resonator absorption behavior and that the larger the pores, the lower the resonant frequency.

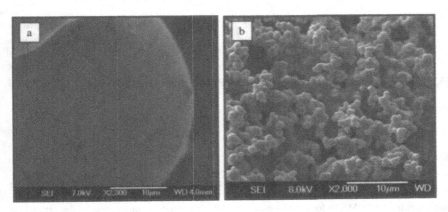

Figure 4. SEMs at the 10 μm-range of ormosil aerogels with a) 40% PDMS and b) 20% PDMS.

In terms of the amount of acoustic energy absorbed, the absorption coefficient also varies as a function of PDMS concentration. (Figure 5) The higher the concentration of PDMS, the higher the level of absorption. This can be explained by the tuning of elastic moduli of the ormosil aerogel with the addition of PDMS. An increased concentration of PDMS results in lower elastic modulus, which results in higher levels of acoustic absorption.

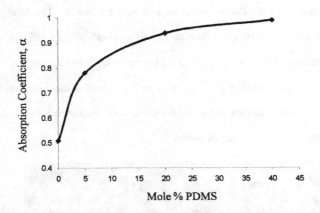

Figure 5. Peak absorption coefficient of silica/PDMS ormosil aerogels as a function PDMS concentration.

Conclusions

The elastic moduli of silica aerogels can be modified and lowered through the incorporation of PDMS to form an ormosil aerogel. The mode of absorption of the aerogels is different than that of commercially available acoustic insulators. Whereas commercially available material exhibited porous open-cell absorption, the aerogels' behavior resembled that of resonators. The moduli and the pore size of the resultant aerogel are functions of the concentration of PDMS, the higher the concentration of PDMS, the smaller the average pore diameter and the lower the elastic modulus. The resonant frequency is generally considered a a function of cavity size, higher resonant frequencies are correlated to smaller pore sizes, which in this case is a higher concentration of PDMS. The level of absorption is a function of the elastic modulus of the material where the lower the modulus (higher concentration of PDMS) the higher the absorption.

Acknowledgement

The authors gratefully acknowledge funding by the Paul S. Veneklasen Research Foundation and the Department of Education (College Cost Reduction and Accessibility Act). Special thank you to Elizabeth Scott and Ulus Ekerman for their contribution to this project.

References

1. Forest, L., V. Gibiat, and A. Hooley, *Impedance matching and acoustic absorption in granular layers of silica aerogels.* Journal of Non-Crystalline Solids, 2001. **285**: p. 230-235.
2. Mackenzie, J., Q. Huang, and T. Iwamoto, *Mechanical properties of ormosils.* Journal of Sol-Gel Science and Technology, 1996. **7**: p. 151-161.
3. Allard, J., *Propagation of sound in porous media.* 1 ed. 1993: Springer.
4. Caponi, S., et al., *Acoustic attenuation in silica porous systems.* Journal of Non-Crystalline Solids, 2003. **322**: p. 29-34.
5. Lee, Y., H. Sun, and X. Guo, *Effects of the panel and helmholtz resonances on a micro-perforated absorber.* Int. J. of Appl. Math and Mech., 2005. **4**: p. 49-54.
6. Allard, J., et al., *Inhomogeneous Biot waves in layered media.* Journal of Applied Physics, 1989. **66**: p. 2278-2286.
7. Sakamoto, S., M. Hikari, and T. Hideki, *Numerical study on sound absorption characteristics of resonance-type brick/block walls.* J. Acoust. Soc. Jpn. (E), 2000. **21**: p. 9-15.

Mater. Res. Soc. Symp. Proc. Vol. 1188 © 2009 Materials Research Society   1188-LL07-05

Nanoparticle Based Multilayers as Multifunctional Optical Coatings

Silvia Colodrero, Mauricio E. Calvo, Olalla Sánchez Sobrado and Hernán Míguez

Instituto de Ciencia de Materiales de Sevilla, Consejo Superior de Investigaciones Científicas (CSIC), Américo Vespucio 49, 41092 Sevilla, Spain

ABSTRACT

Herein we introduce nanoparticle based periodic multilayers as base materials to create different types of multifunctional coatings that combine optical, mechanical and diffusion properties. The technological potential of these versatile materials is demonstrated by showing applications in the fields of sensing and photovoltaic materials. Due to the porous nature of such structures, liquids and gases can infiltrate or condensate, respectively within the interstices, causing a variation of the refractive index (R.I.)of the layers. This gives rise to clear but gradual changes of the optical responses, either when liquids or the partial pressure of vapors are infiltrated in the structure. Also, photoconducting Bragg mirrors can be built by precise control of the spatial variation of the R.I. of the layers in a pure TiO_2 multilayer. Rationally placed within a Dye Sensitized Solar Cell (DSSC), that gives rise to a significant enhancement of the solar to electric power conversion efficiency through the amplification of sunlight absorption. Direct observation of both optical absorption and photocurrent resonances can be seen.

INTRODUCTION

The development of thin films multifunctional materials is one of the most exciting challenges in current materials science.[1] Among them, metal oxide layers have attracted a great deal of attention due to the wide variety of fields in which they are applied and to their relative low cost. The possibility of obtaining particles of these oxides in the range of nanometers has made them irreplaceable compounds in fields such as sensing,[2,3] catalysis,[4,5] and energy conversion.[6] On the other hand, the development of photonic structures has gained considerably attention in the recent years due to the possibility of controlling the flow of light through a material. Particularly, one dimensional photonic crystals (1DPC) are structures that present a periodic modulation of the refractive index in one direction of the space.[7]

In this work, we present a method to create mesostructured Bragg reflectors in which the building blocks are nanoparticles of different sort. The material that we have built has a multilayer structure in which each layer is formed by packed nanoparticles of different composition uniformly deposited by spin-coating of colloidal suspensions of controlled aggregation state. Multifunctionallity in these high optical quality 1DPCs is brought by the combination of optical properties tailored in the UV-VIS region of the spectrum, their large and highly accessible interconnected mesoscopic porosity and the intrinsic properties of each oxide of the structure (electronic transport, surface chemistry, etc). Therefore, we show that all of these features make porous 1DPCs versatile and multifunctional materials which can be used to sense, to integrate in photoelectric devices or to build flexible hybrid organic-inorganic multilayers.

EXPERIMENTAL

Preparation of Suspensions

TiO_2 nanoparticulated sols were synthesized using a procedure based on the hydrolysis of titanium tetraisopropoxide ($Ti(OCH_2CH_2CH_3)_4$, 97% *Aldrich*, abbrev. TTIP) as it has been described before.[8,9] Briefly, TTIP was added to Milli-Q water. The white precipitate was filtered and washed several times with distilled water. The resultant solid was peptized in an oven at 120° C for 3 hours with tetramethylammonium hydroxide (*Fluka*). Finally, the suspension obtained was centrifuged at 14.000 rpm for 10 minutes. Other TiO_2 sol was obtained by peptization of TTIP in an aqueous solution of tetraethylammonium hydroxide solution (35% Aldrich, abbrev TEAOH) at 85°C for 15 hour in a closed vessel. SiO_2 nanocolloids were purchased from Dupont (LUDOX TMA, *Aldrich*). In order to improve adherence, sols were diluted in a mixture 4:1 vol. methanol:H_2O.

Multilayer assembly

Suspensions were spread onto cleaned substrates (normal glass, conductive glass, silicon, etc) and spun immediately at constant velocity using a Laurell WS-400E-6NPP Spin Coater operating at atmospheric pressure. Thermal treatment can be applied to consolidate the structure (5°C/min, 450°C, 30 minutes). Then, a second layer made with other kind of nanoparticles (either SiO_2 or TiO_2) was deposited using the procedure described above. By repeating this cycle a few times, a periodic multilayer displaying 1DPC properties is built up. A final sintering can be applied depending on the use of the multilayer.

Structural Characterization

Particle distribution size of the colloids precursors (SiO_2 and TiO_2 sols) were measured using dynamic light scattering photocorrelation spectroscopy (Malvern Zetamaster S). FESEM images of the multilayers films were taken by using a microscope Hitachi 5200 operating at 5 kV. HRTEM images were obtained with a Philips CM 200 electron microscope. In cross-sectional observations (XTEM), the specimen was prepared by mechanical thinning followed by Ar^+ ion milling.

Optical Reflectance Measurements

Reflectance spectra were performed using a Fourier Transform infrared spectrophotometer (Bruker IFS-66 FTIR) attached to a microscope and operating in reflection mode with a 4X objective with 0.1 of numerical aperture (light cone angle ±5.7°).

DISCUSSION

1DPC Structure

In Figure 1a we show a cross section image of a multilayer obtained by field emission scanning electron microscopy (FESEM). The different morphology of SiO_2 and TiO_2 nanoparticles employed allows to clearly distinguishing the two different kinds of layers in the periodic arrangement. SiO_2 Ludox® particles are spherical and larger than the irregular TiO_2 crystallites. As a consequence of this built up periodicity, the main sign of the interference effects that occur, is a reflectance peak (commonly named Bragg peak) whose position and width depends on the refractive index and thickness of the layers as well as, on a second order approximation, the number of unit cells. The spectral position of the Bragg peak can be tuned by changing the lattice parameter of the periodic structure which is realized by varying the thickness of each type of layer in the structure. Thickness of the deposited nanoparticle films was controlled by changing the concentration of the suspensions (Figure 1b) [10] or the parameters of the spin coater: acceleration and final speed (Figure 1c). [11] By fitting the optical reflectance spectra using a scalar wave approximation,[12,13] we estimated the refractive indexes of the TiO_2 and SiO_2 nanoparticle based films to be n_{TiO2}=1.74 and n_{SiO2}=1.24 respectively, which implies the pore volume fraction in the films, as calculated using Brugemann equation,[14] is 46% in both cases, assuming that bulk TiO_2 and SiO_2 have a refractive index of 2.4 and 1.45 respectively.

The optical response of multilayers can also be modified through the disruption of the periodicity which leads to localized photonic states in the gap. For our 1DPC, the disruption of the periodicity can be achieved by depositing a layer of different thickness in the middle of the stack. FESEM image of a SiO_2 nanoparticle defect built in the centre of a TiO_2-SiO_2 multilayer structure is presented in Figure 2(a). In Figure 2(b) we show the reflectance spectrum of this system. In this reflectance spectrum we can observe one sharp dip in the regions of higher transmittance, within the forbidden band gap frequency range.

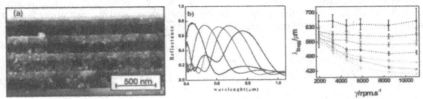

Figure 1. (a) FESEM image of the cross section of an eight layered Bragg reflector made of SiO_2 and TiO_2 nanoparticles deposited alternately on the substrate. The concentration employed in both suspensions is 5% wt. and the rotation speed is set at 6000 rpm. (b) Evolution of the specular reflectance spectra with the concentration of the nanoparticle suspensions the layers are made of. (c) Bragg peak spectral position (nm) versus ramp stage acceleration γ. Colour lines correspond to different final rotation speeds ω (rpm) from 2000 (upper curve) to 8000 (bottom curve) (Reprinted with permission from ref. 10 and 11Copyright 2008. American Chemical Society)

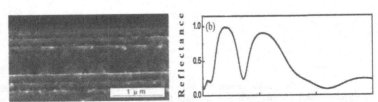

Figure 2. (a) FESEM image showing the cross section of a multilayer in which a SiO_2 nanoparticle defect layer almost 5 times thicker than those forming the periodic structure that surrounds it has been built. The optical cavity is built in the centre of a stack made from silica (3% wt. precursor solution) and titania (5% wt. precursor solution) nanoparticles. b) Optical response attained from the multilayer containing the defect built within the structure shown in (a). (Reprinted with permission from ref. 10 Copyright 2008. American Chemical Society)

1DPCs as base materials for sensing

Besides the structural and optical characterization provided, we analyze the potential of the nanoparticle based 1DPC as base materials for optical chemical sensing devices. The large and highly accessible interconnected porosity present in the multilayer stacks provides a quick and sensitive optical response to different environmental compounds. When an analyte is introduced into the pores of such structures, a variation in the effective refraction index of each layer is produced, which leads to a shift of the Bragg peak that causes a significant change in the colour of the multilayer stack. In this case, we have studied the reflectance peak shifts induced by infiltrating solvents with different refractive index or vapours at different pressures within a multilayer structure of the type shown in Figure 2. We observe that the higher the refractive index of the guest, the larger the shift of the minimum peak (Figure 3a). The dotted line corresponds to reflectance spectra of the multilayer structures in air. The shift of the spectral position of the reflectance minimum of the multilayer is plotted as a function of the refractive index of the different analytes in Figure 3(b). As seen in this figure, the energy variation calculated exhibits a linear correlation with the refractive indexes of the solvents.

.As in the previous case, isopropanol vapours at different pressures also shift the position of the minimum of the spectrum to higher wavelengths due to the condensation within the pores. (Figure 3c) [15]. As we plot the energy variation calculation against the partial pressure, we obtain a vapour adsorption isotherm.(Figure 3d)

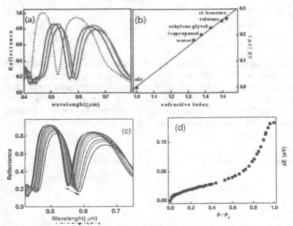

Figure 3. (a) Optical response of the multilayer structure with a planar defect after being infiltrated with the with different refractive index solvents such as water (blue line), ethylene glycol (green line) and Cl-benzene (red line). The dotted line corresponds to reflectance spectrum of the same multilayer structure in air. (b) The shift of the position of the reflectance dip corresponding to a defect state within the bandgap for the different solvent infiltrated multilayers. (c) Optical response of the multilayer structure with a planar defect after being exposed to different partial pressures of isopropanol. The arrow indicates the increase in the vapour partial pressure from 0 to 1. (d) Evolution of the position of the reflectance dip corresponding to the defect state within the bandgap for different partial pressures of isopropanol. (Reprinted with permission from ref.10 and 11 Copyright 2008. American Chemical Society)

234

Besides, these results confirm that optical reflectance is a powerful tool to attain information on the sorption properties of complex structures made by stacking films of different porosity and composition. In this regard, it is worth noting that methods so far reported to analyze the porosity of microporous and mesoporous materials are difficult to apply to thin films successfully, with the exception of ellipso-porosimetry.

Photoconducting Bragg mirrors

A porous 1DPC obtained with a single compound to preserve functionality can be achieved controlling precisely the deposition of TiO_2 nanocrystallite layers of alternate porosity. Then, we generate a periodic modulation of the refractive index that provides the film with photonic crystal properties, well defined and intense Bragg reflections being observed in the visible range.[9] At the same time, the inter- and intra-layer nano-connectivity maintain the photoconductive properties of the TiO_2 without altering others properties of the oxide, such as its surface chemistry. A cross section of this structure obtained by FESEM can be seen in Figure 4a while in Figure 4b we present how the spectral response can be tuned in the visible region, changing the concentration of one of the TiO_2 suspensions.

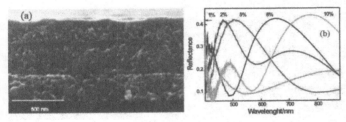

Figure 4. (a) FESEM image of a cross section of a 3-unit cell multilayer deposited on glass. (b) Evolution of the reflectance spectra with lattice parameter. In all cases herein presented, the multilayer structure has been achieved by the alternate stacking of three layers of high packed TiO_2 (concentrations in the precursor solution ranging from 1% to 8% wt.) and two of low packed TiO_2 (5% wt. concentration in the precursor solution). (Reprinted with permission from ref. 9 Copyright 2008 Wiley-VCH)

The combination of optical, adsorption and photoconductive properties of these multi-functional periodical nanostructures was analyzed by photocurrent measurements in different configurations. In one case, the electrodes were made of plane TiO_2 multilayers and their photoconducting properties in the UV region measured to show that the Bragg mirrors herein described are actually photoconducting. In another, the Bragg mirrors were sensitized with a ruthenium dye, commonly employed in Grätzel cells [16] in order to shift the spectral photoelectric response to the center of the visible spectrum

Integration in Dye Sensitised Solar Cells (DSSC)

In a previous work of our group we demonstrated theoretically that the solar-to-electric power conversion efficiency of dye sensitized solar cells can be greatly enhanced by integrating a porous and highly reflecting photonic crystal in the device. [13] Briefly, the light harvesting enhancement is based on the enlargement of optical absorption caused by longer matter-radiation interaction time, which takes place at certain ranges of wavelengths. Photons are localized within the dye-sensitized electrode due to the effect of the photonic crystal, so the probability of optical absorption, and therefore the photogenerated current, is enhanced.

Our 1DPC are advantageous in the sense that they can be integrated easily in DSSC due to their facile method deposition onto the TiO_2 active layer preserving the optical quality. In addition, porosity allows to sensitise the deep layer with the dye and the diffusion of electrolytes are not prevented across the 1DP. Our results in terms of characterization of these DSSC by current-voltage curve are shown in Figure 5. The efficiency improvement in 1DPC based solar cells depends on the thickness of the TiO_2 active layer being larger in thinner (350nm) (Figure 5a) [17] than in thicker (7 µm) electrodes (Figure 5b), [18] maintaining transparency in both cases.

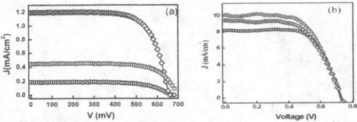

Figure 5. (a) Current voltage curves under 1 sun illumination (100 mW/cm^2) of a 350nm thick dye sensitised nc-TiO_2 electrode coupled to different 1D photonic crystals. The lattice parameter in each case is 140±10nm (blue diamonds) and 180±10nm (green squares). In all cases, the TiO_2 layer thickness is around 85±5nm. The IV curve of a reference cell with the same electrode thickness is also plotted (black circles). (b) Current voltage characteristics for a 7.5 µm thick dye-sensitized nc-TiO_2 electrode coupled to different 1D photonic crystals under one sun illumination. The lattice parameter in each case is 120±10nm (open green triangles), 160±10nm (open red diamonds). The IV curve for a reference cell with the same nc-TiO_2 electrode thickness is also plotted (open black circles). (Reprinted with permission from ref. 17 and 18. Copyright 2009. American Chemical Society and Wiley-VCH)

Flexible Bragg Mirrors

1DPC porous structure can be easily infiltrated with liquids or vapors as mentioned before, but they also can be used as templates of polymers. The infiltration of polymers solutions has been proved as a useful method to fill the pores [19], resulting a hybrid structure which can be lifted-off from the substrate preserving the optical properties. In addition, we obtain flexible multilayers which are capable to be transferred to another substrate. Figure 6a show a cross section image of a SiO_2 - TiO_2 multilayer after to be infiltrated by a polymer and lifted-off from the substrate. In figure 6b we show a macroscopic image of one of this flexible multilayer.

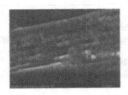

Figure 6 Field emission scanning electron microscopy images of a nanoparticle based 1DPC film infiltrated with poly(bisphenol-A-carbonate). (b) A picture taken while peeling off a polycarbonate infiltrated 1DPC previously deposited on glass. The flexibility of the resulting film can be readily appreciated (Reprinted with permission from ref. 19 Copyright 2009. Royal Society of Chemistry)

CONCLUSIONS

We have shown that nanoparticle based 1DPCs are truly multifunctional materials that could find applications in fields as diverse as sensing or photovoltaics. Such properties arise from the combination of structural, optical and physico-chemical features of the layers, which can be precisely tuned through the preparation conditions and post-treatment.

ACKNOWLEDGMENTS

We thank the Spanish Ministry of Science and Innovation for funding provided under grants MAT2008-02166 and CONSOLIDER HOPE CSD2007-00007, as well as Junta de Andalucía for grant FQM3579.

REFERENCES

1. J.H. Fendler, F.C. Meldrum, *Adv. Mater.* **7**, 607 (1995)
2. C. Garzella, E. Comini, E. Tempesti E, C. Frigeri, G. Sberveglieri, *Sensors and Actuators B-Chemical* **68**, 189 (2000)
3. C. Nayral, T. Ould-Ely, A. Maisonnat, B. Chaudret, P.Fau, L. Lescouzeres, A. Peyre-Lavigne, *Adv. Mater.* **11**, 61 (1999)
4. V.E. Henrich, P.A. Cox, in *The Surface Science of Metal Oxides* Cambridge University Press, 1996, New York
5. H.J. Freund , B. Dillmann, O. Seiferth, G. Klivenyi, M. Bender, D. Ehrlich , I. Hemmerich, D. Cappus, *Catal. Tod.* **32**, 1 (1996)
6. B. O. Regan B, M. Graëtzel, *Nature* **737**, *353* (1991)
7. J.D. Joannopoulos, R.D. Meade, J.N. Winn, *Photonic Crystals: Molding the Flow of Light*, Princeton University Press, Princeton, 1995
8. S.D. Burnside, V. Shklover, C. Barbé, P.Comte, F. Arendse, K. Brooks, M. Grätzel, *Chem. Mater.* **10**, 2419 (1998)
9. M.E. Calvo, S. Colodrero, T.C. Rojas, M. Ocaña, J.A. Anta, H. Míguez, *Adv. Func. Mater.* **18**, 2708 (2008)
10. S. Colodrero, ; M. Ocaña, M.; Miguez, H., *Langmuir* **24**, 4430 (2008)
11. M.E. Calvo, O. Sánchez-Sobrado, S. Colodrero, H. Míguez *Langmuir* **25**, 2443 (2009)
12. K.W.K. Shung, Y.C. Tsai, *Phys. Rev. B* **48**, 11265 (1993)
13. Mihi, A.; Míguez, H. *J. Phys. Chem. B* **109**, 15968 (2005)
14. H.C. Van de Hulst, *Light Scattering by Small Particles*, Dover Publications, ISBN 0486642283 (1981)
15. S. Colodrero S., M. Ocaña, A.R. Gonzalez-Elipe, H. Miguez, *Langmuir* **24**, 9135 (2008)

16. M. Grätzel, *Nature* **414**, 338 (2001)
17. S. Colodrero, A. Mihi, J. A. Anta, M. Ocaña, H. Míguez J. Phys Chem C. **113**, 1150 (2009)
18. S. Colodrero, A. Mihi, L. Häggman, M. Ocaña, G. Boschloo, A. Hagfeldt, H: Míguez, *Advanced Materials* **21**, 764 (2009)
19. M.E. Calvo, O. Sánchez-Sobrado, G. Lozano, H. Míguez J. Mat. Chem. **19**, 3144 (2009)

Printed in the United States
By Bookmasters